W0106555

ISBN 978-94-017-5666-2   ISBN 978-94-017-5956-4 (eBook)
DOI 10.1007/978-94-017-5956-4

THE GENETICS OF THE PIG

by

A. D. Buchanan Smith, O. J. Robison and D. M. Bryant
University of Edinburgh, Institute of Animal Genetics
(*Received for publication May 15th, 1935*)

CONTENTS

The authors desire to record their appreciation of the assistance so freely given by the Staff of the Imperial Bureau of Animal Genetics in general and Miss M. V. CYTOVICH in particular. To the Director of the Institute of Animal Genetics, Professor F. A. E. CREW, they are greatly indebted for his helpful criticism, advice and encouragement.

In the elucidation of doubtful points we gratefully acknowledge the ready co-operation of many of those authors whose work is quoted in these pages. Certain portions of the text have been read by: Professor C. KRONACHER, Berlin-Dahlem, Dr. C. KOSSWIG, Münster, Professor A. L. ANDERSON and Professor J. L. LUSH, Iowa State College, and Mr. H. R. DAVIDSON, England.

The plates illustrate pigs of the different breeds referred to in the text. So far as possible, the photographs represent animals at the age of slaughter. Many gentlemen have assisted us to obtain these pictures; their names are attached to the plates and we take this opportunity of acknowledging their assistance.

I. CHROMOSOME NUMBER

The chromosome number of the pig has been investigated by very few workers. WODSEDALEK (1913) found it to be 18 in both the somatic cells and the spermatogonia of the male, and 20 in the somatic cells and oögonia of the female. HANCE (1917—18) found the diploid number in spermatogonia to be 40 in all cases, with 20 as the number in the first spermatocytes. He obtained variable results with somatic cells, the chromosome number varying from 40 to 58, with one case of 74; this variation may be attributed to fragmentation of the normal chromosomes. His material was taken from animals representing the Berkshire, Jersey-Red and Poland-China breeds. HANCE discusses the results of WODSEDALEK and suggests that either the method of fixing or the use of material which was dead when preserved, caused clumping of the chromosomes, rendering a true count impossible.

KRALLINGER (1931) reports that in the Berkshire and Landschwein breeds the diploid number of chromosomes in the spermatogonia is 38, the first and second maturation divisions each showing 19 chromosomes. BRYDEN (1933), on examination of material obtained from boars of the Large White breed, also found the diploid chromosome number to be 38 and the haploid number 19. With regard to sex-chromosomes, KRALLINGER states that in all diploid plates a peculiar horse-shoe shaped chromosome occurs which is altogether unlike the other chromosomes in the complex. He suggests that this is an X chromosome; although he failed to find a Y chromosome, he considers that, in all probability, this is represented by one of the small chromosomes in the complex. BRYDEN also assumes that the sex-chromosomes are of the XY type.

KRALLINGER, discussing the difference between his own results and those of HANCE, accepts the accuracy of the latter's work and concludes that there are two types of chromosome complement in the domesticated pig, showing diploid numbers of 38 and 40 respectively. He points out that this view is not in conflict with the law of specific chromosome number, since the domesticated pig is the result of a cross between two species, *Sus scrofa* and *Sus vittatus*. It is generally accepted that most of our modern breeds are descended from *Sus scrofa*, with a variable amount of crossing to *Sus vittatus*, though breeds of the pork or lard type, such as the Middle White or the Berkshire, would appear to be derived more from *Sus vittatus* than from *Sus scrofa*. In any case, both HANCE and KRALLINGER obtained some of their material from Berkshire pigs and reported in all cases chromosome complexes differing in number by two. If KRALLINGER's hypothesis that two types of chromosome complex occur in domesticated pigs be correct, it would be expected that all animals of a well established breed would possess the same type. In view of this difference between the results of HANCE on the one hand and those of KRALLINGER and BRYDEN on the other, the present authors consider that the question of the exact number of chromosomes must be left open for further investigation. In the meantime, as a working hypothesis, 19 and 38 might be accepted as approximately correct for the haploid and diploid chromosome numbers in the domesticated pig.

II. COLOUR

1. *General*

Many, though not all, types of wild pigs are of the agouti colour, so commonly found in mammals. This colour is due to varying shades of pigment, from yellow to black, forming a banding pattern on each individual hair. The young of such pigs are frequently (and possibly always) born with longitudinal stripes.

In the domesticated varieties, three fundamental colours are found, black, red and white. So far as is known, the individual hair is always of one colour. (TEODOREANU (1922), however, does not confirm this). Any one of these colours may be combined with any other, or all three may appear in one animal. The distribution of the colours varies from markings which may cover the „points" („acra", i.e. feet, tail and snout) or extend over half the body, to spotting and striping, both longitudinal and transverse. The hairs may be a different colour from the skin: a pig with a black skin and white hairs is usually designated „blue". The reverse, coloured hairs on a white skin, has not been reported. Roaning also occurs as an intimate mingling of white and pigmented hairs. Where the pigmented hairs are black, the description is „blue roan" or „blue grey"; if the coloured hairs are red, the description is „red roan".

There are, of course, modifications of the basal colours. Black may have a red tinge, particularly in a certain light; and some authorities distinguish „jet black". White may be „rich", with a yellow tinge, „chalky", owing to an absence of pigment, or „dirty" which is self-explanatory. The greatest range is in red which varies in intensity from a deep rich dark colour to a yellow cream. As a rule it is not difficult to distinguish the basal colours. It is possible that the „sepia" colour reported by WRIGHT (1918) and MCPHEE and ZELLER (1925) is not a variation of red but a fundamental colour, as in the guinea pig.

Special mention must be made of „tigering" since this term is much employed by investigators on the Continent of Europe. As we understand it, the term was primarily used to describe a red pig with black markings resembling those of a tiger. It has been extended to cover white markings and is now frequently used to describe any tri-coloured pig. „Dalmatian tigering" implies a white pig with black spots, but as the word „Dalmatian" is frequently omitted by certain writers, the connotation of „tigering" has now become extensive. Another type of tigering is that termed „juvenile tigering", observed in wild pigs.

In that the basal colours appear to be more easily distinguished, the problem would seem simpler than in the horse. And further, since the data are drawn not from herd book records, but from direct observations of competent observers, an analysis of the inheritance of coat colour in the pig should be much easier. This, however, is not the case. There exists a multiplicity of breeds of pigs which contain the blood of diverse strains from all over the world. The genotype governing the same phenotype varies radically from breed to breed and even within a breed. In the mating of pigs of different breeds, (unless such a mating has been made before), it is not possible to predict with any high degree of certainty the colour of the progeny.

The purpose of this paper can be served best if the results of the crosses between established breeds are first described. Thereafter, we can deal with some of the principal theories concerning the mode of inheritance of the various colour characters. Except where a special note is made, the colours refer primarily to the hair and not to the skin. Also, except where special note is made, the results of reciprocal crosses, so far as can be ascertained, are the same.

2. *Basal Colour*

a) W h i t e × B l a c k

Large White × Large Black gives, as a rule, white progeny with blue spots (i.e. white hairs on black skin). In certain matings, a proportion (or all) of the progeny has black spots instead of blue. Despite considerable experience, the writers have never encountered an animal resulting from this cross which did not possess at least one black or blue spot.

Middle White × Large Black also gives white with black spots or white with blue spots (as above). Both types may appear from the same mating. Black with white points (like the Berkshire) is also found, though not commonly.

American white breeds × Hampshire (black with white saddle). LLOYD-JONES and EVVARD (1919) state that the progeny is blue roan, i.e. a mixture of black and white hairs growing out of a black skin, except in the area covered by the belt, where both hair and skin are white.

German Edelschwein (Large White) × Cornwall (Large Black). According to KRONACHER (1924), the F_1 of this cross is intermediate, i.e. blue-grey; the colour may range from white with only a few lightly pigmented bristles, to a uniform dark blue-grey. Amongst the F_2 are to be found the following: white, black, white with blue spots, white with black spots (rare), black with red markings, and black tigered with typical red-brown hair on body and legs (rare). The backcross of the F_1 to Edelschwein gives white, and white with blue spots, while with the Cornwall the result is black, and white with blue spots.

German Improved Landschwein (white) × *Cornwall.* KRONACHER (1924) obtained chiefly blue roans (with or without longitudinal stripes) but also black with white saddle, black with Berkshire markings, black and white, white with black spots, and white with blue spots. The black was always deep and glossy. The backcross to the Cornwall boar gave self white, „half black", self black, „tiger" (white, grey and black), and white-and-black. Three matings of three F_1 sows to a Cornwall boar gave litters where all the piglings were self black, save for white markings on the feet and tips of the tail of a few of the animals.

Mangaliţa (dirty white) × *Cornwall.* CONSTANTINESCU (1928) reports black progeny as but points out that the white of the Mangaliţa differs from that of the Large White and German Edelschwein in that unlike the latter, the skin is pigmented. TEODOREANU (1922) states that in the hair fibres pigment is accumulated at the tip and root, (cf. the wild pig in which pigment occurs usually in the middle region of the fibre).

Mangaliţa × *Berkshire (black with white points)* or × *Cornwall.* According to TEODOREANU (1929), the white of the Mangaliţa is incompletely dominant to the Berkshire and recessive to the Cornwall. However, KOSSWIG & OSSENT (1931) found that when the Mangaliţa was crossed with black animals that do not carry the factor for wild colour (*e.g.* Cornwall) or with the Berkshire, all the F_1 was always wild-coloured; this shows that the Mangaliţa carry cryptomerically the factor for wild colour.

Large White × *Berkshire.* The F_1 gives self white pigs. There is no reliable evidence concerning the F_2 or backcrosses.

American white breeds × *Berkshire* or × *Poland-China (black with white points).* According to SMITH (1913), this cross gives white progeny. LLOYD JONES and EVVARD (1919) report self white and white with black spots. REED and CHAPMAN (1927) find the progeny to be white and only occasionally spotted. SEVERSON (1917) states that white with black spots occurs rarely when the parent white breed is Yorkshire (Large White) but quite commonly when it is Chester White. Prof. LUSH, of Iowa State College, in a letter to the present authors, makes the following observation on this cross:

„So far as I am acquainted with these crosses, the results are almost unanimously white pigs, usually with at least a few black spots in the skin but not in the hair. Occasionally there is a purebred Yorkshire or Chester White boar which is not homozygous for the dominant white gene, and in such cases very nearly half of the progeny will show colour of one kind or another. It seems to be a safe conclusion from the general, though poorly controlled, observations of breeders that heterozygosity for the dominant white is rarer in the Yorkshire than in the Chester White. However, heterozygous individuals are rare, even in Chester Whites". From one such mating, using a Yorkshire boar, Professor LUSH also obtained a pig white in its fore parts but distinctly red in the hind quarters; there might have been some doubt of its paternity. WARWICK and BELL (1930) and WARWICK (1931) comment upon the mode of inheritance of the black of the Poland-China.

German Edelschwein × Berkshire. According to NACHTSHEIM (1922), this gives self whites, though spotted progeny are also reported. KRONACHER (1924 and 1930) confirms this: he obtained some red in the F_2 of this cross.

German Improved Landschwein × Berkshire (white). KRONACHER (1924) finds the progeny of this cross to be invariably white.

b) W h i t e × R e d

Yorkshire (Large White) × Tamworth (red). SEVERSON (1917) observed that the progeny was white with reddish tints, though white with black spots occurred rarely. SPILLMAN (1906) had previously stated that the results of this cross were pure white but with a dark skin which showed white spots, and a white belt. An F_1 male × Tamworth sow gave 4 Tamworth red and 4 light grey tinged with red. SPILLMAN's work indicates that there are two types of self red, as judged by the reactions with white. The evidence of REED and CHAPMAN (1927) confirms that the F_1 of the Yorkshire and Tamworth are usually white; this is supported by FROST (1933), who obtained all white animals. SIMPSON (1911) found that roans regularly occurred as a result of backcrossing to Tamworth.

Middle White × Tamworth. KRONACHER (1930) reports an F_1 of 24, all white, except one which had a few red hairs on the sacrum and forehead and another with two small red spots over the eyes. Later (1932), from among 37 piglings from the same cross, 26 were pure white while 11 were white tigered with more or less pronounced red colouring on definite skin areas such as ears, forehead, nape of neck, ribs, back etc., which were clearly visible under the hair. In the F_2, the ratio of 61 non-tigered: 22 tigered was observed. The observations of FROST (1933) on a similar cross are in accordance with this.

Edelschwein (white) × *Tamworth.* KRONACHER (1930) obtained two litters, the Tamworth sows being the same as those used for the preceding experiment. In the two litters, two males and two females were pure white; the remainder had some few red hairs, some of these being „intense" in shade. Dark pigmented spots in the skin were also seen in some of the animals. Similar results were reported in 1932.

Edelschwein × *Half-red Bavarian Landschwein.* KRONACHER (1924) reports the majority of offspring from this cross to be white, but an appreciable proportion of white animals with black spots also resulted. In another series of matings observed by the same author, several hundreds of such animals were produced and all were pure white save five showing red spots. In the F_2 and F_3, the following were obtained: white, white with black spots, red (with and without white markings), and red with black. The backcrosses of the F_1 to the Half-red produced white, red (mostly $^3/_4$ red), red with black, and white with reddish stripes. In the adult coat some of these animals exhibited a light grey-black tinge in their red coat. The backcrosses to the Edelschwein invariably gave white. Wild striping appeared as the result of mating two animals produced from the same litter of F_1 × Half-red.

Improved Landschwein (white) × *Half-red Bavarian Landschwein.* KRONACHER (1924) only records the mating of Half-red sows and Landschwein boars. The F_1 were all white except one animal which had large pigmented spots on the ear and on the hind-quarters. The pigment had a reddish tinge, the hair on the pigmented areas being partly white and partly light red. Among the F_2 of 50 animals, there were produced 10 piglings with black spots and one with light-red hindquarters bearing large black ticks. Others were produced with black spots, but the majority were like their white parents. In the backcross to the red breed, one Bavarian sow produced a litter of 6 white: 7 pigmented, while three F_1 litter sisters × white boars gave almost exclusively whites. In one such mating, six were white, while four had blue spots: in an identical mating the litter sisters produced only whites.

Mangaliţa × *Tamworth.* TEODOREANU (1929) reports the so-called white of this Rumanian pig (which is found throughout southeast Europe) as intermediate when crossed with the red Tamworth. KOSSWIG & OSSENT (1931) state that the Mangaliţa always breed true but that the F_1 of Mangaliţa by any colour always gives wild-coloured progeny.

c) R e d × B l a c k

Tamworth (red) × *Hampshire (black with white saddle).* FRÖLICH (1913) reports the black and white of the Hampshire as dominant to the red of the Tamworth.

Duroc-Jersey (red) × *Hampshire.* LLOYD JONES and EVVARD (1919), both from observation and planned experiment, were able to show that the black colour behaves as a simple dominant to red.

Tamworth × *Large Black.* In this cross (FROST 1933), the F_1 were mainly red with black spots but a few animals were wholly red.

Duroc-Jersey × *Mulefoot (black).* DETLEFSON and CARMICHAEL (1921) report that all the F_1 are black. The backcross to the Duroc-Jersey gave 19 black : 23 red. The shade of red varied from a cream which was almost indistinguishable from white, through lemon and yellow to red : these pigs also showed some roaning. One black pig had a white spot on the upper lip. Yellow and lemon mated to yellow gave cream, yellow and red. Two matings of cream × cream gave 8 cream : 3 yellow : 1 red.

Half-red Bavarian Landschwein × *Cornwall (Large Black).* KRON-ACHER (1924) found all the F_1 to be of a glossy black colour with white markings on the face and fore feet. In the black coat of one young pig were seen red-brown stripes (wild stripes) which disappeared with age. In the F_2, the following appeared: black with white markings; black with white markings and white shoulder; white with glossy red hairs, black tufts·on hind quarters and „tigering"; red with or without white markings but always with some small black spots; self black; black with red sheen in certain lights. The backcross of the F_1 boar to a Bavarian sow produced red pigs with white markings, some with, and some without, black spots.

Bavarian Landschwein × *F_1 Cornwall-Berkshire.* KOSSWIG and OSSENT (1931) mention that two black-white-red tigered piglings occurred among the F_1. The F_2 were as follows: 6 white; 1 black-and-white tigered; 2 black with red sheen; 3 red-black-and-white tigered; 8 completely or partially blue-grey.

Bavarian Half-red × *Berkshire*. KOSSWIG and OSSENT obtained an F_1 spotted black and red.

Tamworth × *Berkshire*. WENTWORTH and LUSH (1923) found 10 F_1 of this cross to be red with black spots. Compared to the Duroc-Jersey × Berkshire cross (*vide infra*), the F_1 showed less black marking; in fact, two pigs showed no visible black. According to FROST (1933), in England this cross gives the same result viz., red pigs with black spots.

Duroc-Jersey × *Berkshire*. According to SEVERSON (1917), this cross gives red in several shades, with black spots. This is confirmed by LLOYD JONES and EVVARD (1919). WENTWORTH and LUSH (1923) obtained in the F_1 yellowish red (in most cases, sandy) animals with small black spots scattered irregularly over the body, but somewhat more frequently on the underline and rear parts. In the new-born pig, a few of these spots were more than a square inch in area, while the majority were not more than one third of that size. A fairly large F_2 was raised as follows: 47 black and red, 27 black and white, 14 self red, 11 black and sandy, 10 black, red and white, 10 black, red and sandy, 7 sandy, 7 sandy and white, 6 black, sandy and white, 3 red and sandy, 3 red and white, 3 white. The backcross of the F_1 boar to his own Duroc-Jersey dam gave: 19 black and red, 14 self red, 1 black, red and white, 1 red and sandy.

The same writers report two interesting cases of „juvenile striping" which they observed in both the F_2 and F_3 generations of the Berkshire × Duroc-Jersey cross. Two F_2 piglings, sandy coloured with light bellies, showed longitudinal stripes similar in pattern to those of the young of the wild hog, except that the coloration was not so intense. This striping disappeared within a few weeks after birth. They obtained also, from unstriped F_2 parents, an F_3 litter of 11 piglings, all of which exhibited even more distinctly the striped pattern.

Duroc-Jersey × *Poland-China*. LLOYD JONES and EVVARD (1919) report that the F_1 is the same as when the Berkshire is the black parent. WARWICK (1926a, 1931) confirms this; he obtained an F_2 of 161 red with black spots: 55 red. The backcross to the Duroc-Jersey produced 123 red with black spots: 139 red. Matings were made of the non-black segregates: 129 of the offspring were non-black, while 5 (from 4 different mothers) had definite black spots. Non-black × Duroc-Jersey gave 88 non-black: 2 each with a small black spot.

d) White×White; Red×Red; Black × Black

There appear to be no accurate observations reported of the results of matings between breeds of the same colour. It may be taken that each breed mated to another of the same colour breeds true to that colour. The off-type colours which appear in pure breeds are worth noting. The white breeds occasionally throw animals with blue spots and still more occasionally, with black spots. CONSTANTINESCU (1933) obtained a pure white F_1 as a result of crossing Mangaliţa with Large and Middle Whites. In the F_2, the Mangaliţa white segregated in four cases. Tiger striping was observed in both the pure and Mangaliţa. whites, the ratio of tigering to non-tigering being 9 : 7. Thus tigering apparently occurred in combination with the white of the Mangaliţa. Juvenile tigering appeared in the F_2. KOSSWIG and OSSENT (1931) observed wild-coloured animals from the Mangaliţa × Middle White cross. It is difficult to appreciate the distinction between the Mangaliţa, which is predominantly white, and the blue pig, which has a black skin and white hairs. The results of the F_2 and backcrosses, while not conclusive, are in accordance with the evidence of those in the F_1. Further mention of this cross is made by WALTER in 1929 and by TEODOREANU in 1932. OSSENT (1929) obtained similar results: of 9 pigs, the F_1 of these breeds, three were black and six were wild-coloured.

ERHARD (1902), as quoted by DECHAMBRE (1929), reported the appearance of two striped piglings in a litter from two completely white parents; the remainder of the litter were pure white. The sire was a pure-bred Yorkshire while the dam was a cross Yorkshire × Large Ear, the latter breed being described as having a light coat. After two months the coats of the two piglings became paler.

CONSTANTINESCU (1933) states that juvenile tigering seems to be due to an accumulation of pigment in certain hairs and consequently disappears when these hairs are shed. This striping varies in intensity and distribution. He believes that the white F_1 from Mangaliţa × Large or × Middle White carry the gene for juvenile striping but do not manifest it owing to the lack of pigment upon which it can act. This postulate is supported by the fact that some of the Large White sows throw progeny which have yellow spots and show juvenile striping on all the spotted areas.

McPHEE and ZELLER (1925) report tri-coloured pigs as the result of inbreeding Chester Whites. The red breeds have only once been reported as throwing blacks. WARWICK (1931), a most reliable observer, states that he has seen a number of Duroc-Jerseys and Tamworths which definitely showed a small amount of black; of 274 pure-bred Duroc-Jerseys in one particular herd he observed only one that showed any black and this consisted of only a few black hairs on the poll at birth. When the non-black (red or sandy) segregates in the F_2 of the Poland-China × Duroc-Jersey cross were mated to the red Duroc-Jersey, WARWICK reports that out of 90 offspring, only two showed black, in each case the spot being small. The black breeds occasionally throw some reds and some with considerable white markings. Large Black × Berkshire gives black (CARR-SAUNDERS, 1922). On the other hand, KOSSWIG and OSSENT (1931), from a mating of a boar and sow of the Güstin Pasture breed, obtained an F_1 of black and tigered in the proportion of 31 : 7, which nearly approaches a ratio of 3 : 1.

3. *Wild Colour*

According to FRÖLICH (1913), the white colour of the Edelschwein behaves as a simple dominant to the grey-black of the European wild pig, which he found to be dominant to the red of the Tamworth. This point is confirmed by SPILLMAN (1906) reporting a backcross of two F_1 sows from Wild × Tamworth by a Tamworth boar, in which he obtained six wild and six red piglings. According to SIMPSON and HER-MANN and HENSELER (see FRÖHLICH 1913), the juvenile striping pattern and the wild colour of the old animals are dominant to the red of the Tamworth. In all the 38 pigs of the cross involving one wild parent, raised by WENTWORTH and LUSH (1923), there were „distinct" longitudinal stripes about 1 cm. in width, composed alternately of rather light red and very dark brown hairs. These stripes extended all over the back and sides but the bellies were a uniform light red. The 4 F_2 pigs were similarly but not so regularly striped; one had a grey belly and one a reddish belly with black spots. The bellies of the other two were like those of the F_1. The backcross out of a Tamworth sow gave 5 striped : 3 non-striped. Of the five striped, in three the stripes were faint and accompanied by black spots on the body. Of the non-striped pigs one was a self red and white, and the other two had black spots. The same authors report a cross of Wild × Berkshire which gave 17 pigs like the F_1 described above but with a fainter red tinge and with black spots. These spots were larger than those found by the authors in the crosses involving the Tamworth.

Wild × Improved Landschwein. KRONACHER (1930) reports the mating of a wild boar to six sows of the white Improved Landschwein. In one litter, two piglings of the wild colour were obtained, the remaining six being white; three litters produced only white, while the fifth litter proved to be a veritable tartan including blue grey, red, and brown (wild colour, striped partly with white hairs). All the white pigs from this cross showed decided striping.

Wild striping has been reported as the result of the F_2 of many breeds which have been crossed and also from the short-lived Sapphire breed.

Wild × Mangaliţa. Mangaliţa × wild-coloured animals of mixed ancestry produced, in the herd of KOSSWIG and OSSENT, an F_1 all wild-coloured. The backcross to Mangaliţa gave 12 wild : 3 Mangaliţa. Mangaliţa crossed with wild-coloured which also had black spots gave a wild-coloured F_1, of which 25 were spotted and 20 non-spotted. Mangaliţa mated to wild-coloured non-black-spotted pigs segregated in the F_1 into 50 wild non-spotted: 10 wild spotted.

Wild × Hannover-Braunschweig Landschwein (black with white saddle). HÆCKER, as cited by KOSSWIG and OSSENT (1931), reported that the offspring from this cross were black with the white belt of the Hannover-Braunschweig, while in the F_2 there were 4 black: 2 wild offspring.

Wild × Berkshire. KOSSWIG and OSSENT (1931) state that the wild colour is dominant, the progeny from this cross being all wild-coloured. Some of these were spotted. However, the wild-coloured parent, being the F_2 from a cross of wild by Hannover-Braunschweig, could only be considered as heterozygous. The further mating of 15 wild-coloured sows to one Berkshire boar gave, according to the same writers

(1932), 52 wild : 8 tigered : 51 black. This segregation may also be considered as 52 wild : 59 non-wild.

Wild × Black. Kosswig and Ossent made 19 matings of wild-coloured pigs of mixed ancestry with some „mainly black" pigs and obtained 238 offspring giving a segregation of 190 coloured : 48 tigered. A further suggestion of a 3 : 1 ratio was obtained in other crosses.

European Wild × Chinese Mask (black with white feet). The progeny of this cross was all black (Kosswig and Ossent, 1931).

Wild × Half-red Bavarian. The result of this cross made by Kosswig and Ossent (1931) was two wild-coloured unspotted F_1.

Darwin (1868) notes that the young of wild European and Indian pigs for the first six months are longitudinally banded with light-coloured stripes, and that the Turkish and Westphalian pigs have striped young „whatever may be their hue". In the chapter on „Inheritance", he states that the best known case of reversion and that on which the wide-spread belief in its universality apparently rests, is that of pigs. „These animals have run wild in the West Indies, South America, and the Falkland islands, and have everywhere acquired the dark colour, the thick bristles, and the great tusks of the wild boar; and the young have re-acquired longitudinal stripes".

Some interesting information concerning the reversion of domesti-cated to other colours could probably be gleaned from a study of the wild pigs of New Zealand. According to Thomson (1922), there exist most exact data as to the introduction of the species into New Zea-land. In June 1773, on his famous second voyage, „Captain Furneaux put on shore in Cannibal Cove a boar and two breeding-sows, so that we have reason to hope this country will, in time, be stocked with these animals, if they are not destroyed by the Natives before they become wild, for, afterwards, they will be in no danger". In the following year, it is recorded that none of the pigs could be found and it was concluded that since few Natives came that way, the pigs had retreated into the thickest parts of the woods. In 1773, Captain Cook gave a few pigs to some Natives near Cape Kidnappers. Thus pigs were introduced into both the South and North Islands. The pigs in the South Island set free at Cannibal Cove later got the nickname of „Captain-Cookers". There was a further introduction by Governor King, of New South Wales, to the Bay of Islands in 1793. Presumably these and subsequent importations consisted of domesticated pigs, i.e. pigs of non-wild colour. The increase of the wild pigs in pre-settlement days was remarkable. Nearly every sealing and whaling vessel which visited these Islands between 1800 and 1830 took away quantities of pork. Later the pigs became a pest and their extermination was sometimes contracted for by experienced hunters; it is on record that three men in 20 months on an area of 250,000 acres, killed no fewer than 25,000 pigs.

At the present time wild pigs are still common in nearly all scrub or thin bush country which is not too near a settlement. The wild pigs of the Otago district are reported to have been „originally a variety of the Tamworth breed — long-snouted, razor-backed, built for speed rather than for fattening, quick and agile in movement". Their colour was red or sandy red, with some black, and there were also a few black and white. They did not appear to have reverted to the wild coloration. Other evidence points to the fact that wild coloration is unknown amongst these pigs.

4. White Markings

White Points—Berkshire. According to WENTWORTH and LUSH (1923), the Berkshire crossed to the Tamworth and the Duroc-Jersey gives pigs showing no white. Out of an F_2 numbering 145 (omitting self whites), there were obtained 92 without white: 53 white-spotted. The backcross of the Berkshire-Duroc Jersey boar $\times$ Duroc-Jersey sow produced 35 offspring, one of which showed white markings. CARR-SAUNDERS (1922), mating Large Black boars $\times$ Berkshire sows, obtained self colours, while the reciprocal cross gave in the F_1 a gradation from self black to spotting in which the white coat was evenly divided into black and white patches. The limited experience of the present writers is that the reciprocal cross gives self colours. We have also observed the Berkshire marking as the occasional result of the cross between the Large Black and Middle White. NACHTSHEIM (1922) states that Berkshire $\times$ white usually gives white, rarely spotted, and never black.

In this connection the breed called the Kentucky Red Berkshire, must be mentioned. Further reference to this breed is made on page 28.

Self colour black and red breeds are always liable to throw animals with white spots, especially at the extremities. In several breeds, white feet are not uncommonly found though their occurrence is seldom reported. EVANS (1930), in America, indicates the hereditary nature of this, and the present authors have observed it in Great Britain. NORDBY (1934) has made a study of white spotting in pure-bred Duroc-Jersey swine. This is liable to occur on the extremities and is associated in some cases with a partial or complete saddle. The condition appears to be intensified by inbreeding. Experimental matings were made, both of self-coloured pigs known to have produced white-spotted offspring, and of pigs showing white on the extremities. In all, 196 pigs were obtained in 21 litters, of which 128 were self red and 68 were marked with white on one or more of the feet. In addition, one pig showed a complete saddle and one an incomplete saddle over the shoulders and fore-ribs; one exhibited a saddle over the rump; three showed white tail-tips, and one had a small blaze over the forehead. It is interesting that in son × dam matings, 60 per cent. of the offspring, and in full brother × sister matings, 50 per cent., showed white markings.

Saddle. SPILLMAN (1907b) found the saddle of the Hampshire to be dominant over self colour and this has been confirmed by FRÖLICH (1913), in a cross with wild, SIMPSON (1914), LLOYD JONES and EVVARD (1919), and DURHAM (1921). In crosses between the Hanno-ver-Braunschweig and the wild pig, KOSSWIG & OSSENT (1931) obtained progeny which were black but carried the white belt of the former breed. HOBSON (1931) has produced a breeders' symposium on this subject. The saddle behaves in the same manner as the belt in cattle, though whereas in the latter the belt occurs between the fore and hind legs, in the pig it is found on the shoulder and includes the forelegs. As in cattle, there is great variation in the size of the belt, while in crosses with self breeds it is often considerably diminished. From an historical point of view, it is worth noting that this has been cited as evidence of „gametic impurity". Attention must be drawn to the fact (reported above) that NORDBY (1934) has found in the self-coloured red Duroc-Jersey breed pigs exhibiting saddle amongst a strain showing other white markings.

Half-coloured. KRONACHER (1924) reports matings which in-volved the Half-red Bavarian Landschwein. Such an animal possesses white points, a white underline and a broad white shoulder together with a white face; there were also some with coloured markings round the base of the ears. As the Edelschwein to which they were mated possessed a dominant white, no information could be obtained from the F_1. In the F_2, a few animals were half-coloured, with varying amounts of white on the face, a white saddle, white feet, and black ticking over the whole body. The backcross of the $F_1 \times$ the Half-red dam produced a variety of white markings on the coloured offspring but again no typical Half-red appears to have been recovered. The same author, in the mating of the Cornwall × Half-red, obtained an F_1 of self blacks. Again in the F_2 a variety of white markings was observed while in the backcross to the Half-red, the half-red pattern was obtained. KRONACHER observes that the factor for the „half pattern" acts only upon red.

White Face. The „Hereford Hog" once appeared in the United States of America. A Journal was published in its support and a claim was made that the breed was resistant to swine fever. Prof. LUSH writes: „For the last ten or fifteen years there has been an attempt to start a breed of hogs known as Hereford Hogs centering around the town of La Plata, Missouri. The exact date of origin was some time before 1919. I first saw them in 1921. These pigs are red with white feet and faces and often a white tail, hence the name. The amount of white is distinctly greater than in the standard Berkshire or Poland China marking, and very often pigs are produced which are distinctly red and white spotted with the white on other parts of the body". We believe that specimens of this breed are still to be found in Missouri.

Roaning. As already noted, roans may arise in a variety of ways. SIMPSON (1914) obtained this colour in the backcross of Yorkshire/Tamworth × Tamworth. It appears regularly in the F_1 of the crosses in which the Hampshire is involved with white breeds. McLEAN (1914) reported that the Sapphire Hog, which was a „blue" roan, had a very complex pedigree containing Hampshire, Berkshire, Essex, Chester White and Duroc-Jersey blood; the Tamworth and Poland-China are reported to be quite free from blame in respect of the paternity of this breed, some descendants of which are reported to be still alive. There is some reason to believe that the colour did not breed true. It was in connection with this breed that „Mackerel marking" has been mentioned. Professor A. L. ANDERSON, of Iowa, informs us that the booklet of the Association of Breeders of the Sapphire hog states as follows: „The animals are rich blue or bluish grey in colour, differing in but one very essential feature from any breed ever produced. The colour may be called blue grey, blue roan or iron grey". WENTWORTH and LUSH (1923) point out that as a general rule, roans appear at birth to be self-coloured, the roaning becoming evident at weaning time.

Spots. The spotted pattern of the Spotted Poland-China and Gloucester Old Spots appears to be black spots on a white ground rather than white spots on a black ground. The black spots are reported in the F_2 of a variety of crosses and appear to behave in a recessive manner (See also MALSBURG 1924). Apparently they are the same as the black spots on sandy ground which are characteristic of the Tamworth × Large Black cross, and which have been already described in the Duroc-Jersey × Berkshire and Duroc-Jersey × Poland-China crosses. In this connection, WENTWORTH and LUSH (1923) state that in many cases sandy may be substituted for white. Professor A. L. ANDERSON has kindly sent us a photograph of a Hampshire boar which had been in use in the Iowa College herd. When this boar was a little over two years of age, he developed spotting which increased in intensity until the time he was disposed of some six months later. Professor ANDERSON reports that he has located three other animals of this same breed which turned a similar colour; he states that he has heard of the condition in the Duroc-Jersey breed. Judging from the photograph, the animal developed white spots on the black skin and hair, giving the appearance of large snow flakes on a black ground.

Sepia. McPHEE and ZELLER (1925) describe a sepia coloured animal obtained by inbreeding Poland-Chinas. The colour appeared to be a simple recessive to black. As already noted, sepia may possibly be a distinct and fundamental colour.

Albino. As far as can be ascertained, no albino pigs have ever been reported.

5. Analysis of Results

Hitherto the present writers have dealt with facts as reported by observers. Upon these facts the following observations may be made:

White appears to be genetically the same in all breeds, the only exception being the „dirty white" of the Mangaliţa.

There exists a number of different blacks. The Hampshire, Large Black (Cornwall), Hannover-Braunschweig and the Güstin Pasture do not behave as though of similar genetic constitution. It is difficult to determine how much of this variation is due to heterozygosity for other colours masked by the black. The black of the Berkshire and the Poland-China is definitely distinct from that of the other black breeds.

As regards red, there are almost certainly at least two genotypes for this colour. Probably there are also modifiers which, however, have a lesser effect than those major modifying factors affecting black. On the other hand, the difference between the various red breeds may be due to a heterozygous condition of certain cryptomeric factors. This is probably only another way of writing the previous sentence.

The wild colour can be extracted from matings involving white, black and red parents, either in the F_1 or the F_2, depending probably on the homozygosity of the parent breeds. The wild striping can certainly be so obtained and possibly also the agouti.

The Berkshire and Poland-China markings are probably not white marks on a black body.

The white saddle (or belt) is a simple dominant to self colour. SPILLMAN's (1907a) results, however, demand two pairs of factors and in any case the action of modifying factors which affect the extension of the belt. NORDBY (1934), who obtained white saddled pigs from a self-coloured breed, explains that „the factors for belting. are independent of those for colour. However, the belt cannot appear in the absence of restriction factors for red, as these apparently permit white to come into evidence."

Half pattern behaves in a recessive manner. This is possibly the same as the white face of the Hereford hog.

The following writers deserve particular mention for their work in attempting to resolve this problem of colour inheritance. From America comes the work of SEWALL WRIGHT followed by that of WENTWORTH and LUSH, and from Germany the work of KRONACHER, and subsequently KOSSWIG and OSSENT. The findings of WRIGHT are fully discussed by WENTWORTH and LUSH in their series of papers which embody a great deal of the earlier work. KRONACHER reached conclusions differing from those of WENTWORTH and LUSH principally in detail. On the publication of the findings of KOSSWIG & OSSENT, he re-examined his data (KRONACHER & OGRIZEK 1932) and came to the conclusion that they confirmed those of KOSSWIG & OSSENT. His former conclusions are therefore not described but any future investigator should make a point of studying his exceedingly comprehensive papers which are full of illuminating points.

Accordingly, there are two sets of conclusions which deserve closer study, those of WENTWORTH and LUSH, and those of KOSSWIG and OSSENT. These are therefore summarised. The observations of the present authors will be found in the parallel column.

TABLE I. INHERITANCE OF COLOUR.

WENTWORTH and LUSH	Remarks
Wild Colour.	
1. Factor for immature striping (adult agouti), simple dominant to red and to black.	The presence of two types of wild marking is supported by DECHAMBRE.
2. A recessive intensifying factor.	
Black Spotting hypostatic.	Black spotting can be taken to correspond with what KOSSWIG and OSSENT call tigering.
A factor subject to modifying factors, e.g. Wild × Tamworth (red) frequently gives black-spotted progeny. This factor only manifests itself in certain combinations though it may be present in others.	
The Berkshire and the Poland-China are black-spotted pigs with independent factors extending the black. This behaves as dominant in crosses with red	Investigators are generally agreed on this point.

and as recessive in crosses with self white.

Black, self.
Dominant to red and recessive to white.

Further evidence shows that the position is not quite so simple. This explanation cannot account for all the cases noted.

White, self. ˙
Dominant over all other colours and probably dependent on a single factor.

N.B. In certain crosses with reds and with blacks, roan is obtained.

Except as regards Mangaliţa.

Red, self.
Two kinds genetically distinct in relation to self white. Duroc-Jersey carries two factors which act cumulatively.

The only alternative to this solution would appear to be the same fact stated differently, viz. only one red but it can carry cryptomerically factors affecting other genetic constitutions.

Red Spotting.
Three types.
1. Roan.
2. Irregular spots on lighter ground.
3. Lighter coloured belly.

The white of the Berkshire and Poland-China is a red or sandy diluted by accumulated modifying factors. This white appears to be entirely distinct from the self white and the white saddle of the Hampshire.

The conclusion that Berkshire and Poland-China are really red pigs with black spots is one which practically every competent investigator has reached.

White Saddle.

Dominant over absence but influenced by modifying factors.

There is no dispute on this point.

The conclusions of KOSSWIG & OSSENT (1931), as revised in 1933, are as follows:

Allelomorphic Series	Remarks
Rub_{ep} — Epistatic black e.g. Hannover-Braunschweig	Only a dominant black can account for the colours of the progeny involving the Hannover-Braunschweig. The Hamsphire is probably of the same constitution.
Rub_{hyp} — Hypostatic black e.g. Cornwall (Large Black), Güstin Pasture.	
rub_{ti} — Tigering — i.e. red with black spots, e.g. Berkshire × Poland-China in association with modifying factors. Also an alternative constitution for the self red of Tamworth and Duroc-Jersey, in association with factors for restriction of red. This gene can also account for red-black-and-white-tigered, black-and-white-tigered, red-and-black-tigered, red-and-white-tigered, and white.	In agreement with WRIGHT (1918) and with WENTWORTH and LUSH (1923).
rub — Red. e.g. Half-red Bavarian Landschwein. Also alternative constitution for Tamworth and Duroc-Jersey.	Two possible genotypes for the red phenotype is in agreement with WENTWORTH and LUSH.
Uni — wild $\begin{cases} \text{Linked to } Rub_{ep} \\ \text{and } Rub_{hyp} \end{cases}$	This is required to explain the wild progeny in the F_1 or F_2 of black and red crosses according

uni — not wild $\begin{cases} \text{Linked to } rub_{ti} \\ \text{and } rub \end{cases}$

to whether the black is epistatic or hypostatic.

The linkage is strong and no crossing over has been reported.

In effect wild colour must be carried cryptomerically in a black phenotype but not in a red.

The existence of two types of wild colour is not disputed.

Modifying factors for all the above.

fla — Dirty white of Mangaliţa is caused by a recessive factor, epistatic to all colours except perhaps Rub_{ep}.

This appears reasonable.

Col and $Real$ — Two dominant complementary factors causing the white of all the white breeds except Mangaliţa. There may be more than two of these.

Hom — White belt, simple dominant over absence, subject to modifying factors.

No dispute.

Mon — Half colour, simple recessive to self colour. Anterior half of pig is white.

No dispute.

KOSSWIG & OSSENT draw attention to the fact that their hypothesis is analogous to the mode of inheritance of colour in rodents and mention a paper by HALDANE (1927) in which the colour series of rodents and carnivores are compared. The agouti (A/a series of English writers) is represented by Uni/uni. The E series of English writers is analogous to Rub_{hyp} rub_{ti} .... rub allelomorphic series, while the English C (albino) series is perhaps represented by the sepia reported by MCPHEE & ZELLER (1925).

It is not the purpose of this paper to formulate a definite conclusion unless there is a general concensus of opinion that this conclusion is correct, nor to adumbrate a new theory. There can be no doubt that KOSSWIG & OSSENT have made a good case for their interpretation of the mode of inheritance of colour in the pig. It suffers, however, from the fact that one genotype, rub_{ii} rub_{ii}, covers such a variety of phenotypes, and one phenotype, black, requires a variety of genotypes. There is much to support their conclusions and quite possibly they are correct, but we submit that on many points further investigation is desirable and that until the phenotypes constituting the genotype rub_{ii} rub_{ii} have been reduced to order, it will not be possible to promote the hypothesis to the rank of theory.

There remain yet one or two points which require to be cleared up. As regards black × white, the usual result is that the progeny have blue spots, but occasionally it happens that these spots are black. In cases where black spots appear in the offspring, CREW (1924a) found that, if a sufficient number of matings be considered, blue-spotted and black-spotted offspring occur in equal numbers. He suggests that the blue-spotted condition may be due to the presence of a dominant dilution factor which turns black into blue. When the Large White male (this cross is almost invariably made by mating a Large White boar to Large Black sows) carries this factor in the duplex state, all the F_1 generation are blue-spotted. If, however, the Large White sire carries this factor in the simplex state, some of the F_1 generation are black-spotted. Without definite reason, black-spotted pigs are disliked by English feeders and curers. The assumption at present held by many breeders that these blue-spotted and black-spotted pigs do not breed true has not been adequately verified. It is of some importance to find out whether the black-spotted pigs resulting from the Middle White × Large Black cross can be made to breed true and so become like the Gloucester Old Spots or Spotted Poland-China.

The spotted breeds such as the Spotted Poland-China and the Gloucester Old Spots must be genetically allied to the black breeds with white markings such as the Berkshire and Poland-China. The difference between these breeds, so far as colour is concerned, seems to depend on the action of the different modifying factors. WRIGHT (1918) appears to have been the first to suggest that the Berkshire is a modification of the tortoiseshell (tri-colour) pig, and it is interesting to note how his hypothesis has stood the test and is used by WENTWORTH and LUSH, KRONACHER, and also by KOSSWIG and OSSENT in their interpretations. In this connection, DETLEFSON and CARMICHAEL (1921) state that the yellow pigment which they observed by microscopic examination in the hair of their cream pigs, was also found by them in the white hair of Berkshires.

Roaning is˙probably a modification of the „blue" spots. The data — or rather lack of data — concerning the „Sapphire Hog" confirms this (McLEAN 1914).

In strong support of the hypothesis that the Berkshire and Poland-China black is of a different nature from that of the other black breeds, and is merely an extended form of black spotting, is the fact that the Duroc-Jersey (red) crossed with the Berkshire produces red pigs with black spots, but when it is crossed to a black breed the black colour is dominant. WARWICK (1931) described the results obtained in crossing Poland-China or Berkshire pigs with Duroc-Jerseys or Tamworths. In all, 131 F_1 offspring were produced and all exhibited a certain amount of black. The F_1 did not carry as much black as their black parents, the total amount being approximately half or even less. In body colour, they were red or sandy with black spots of various sizes. There were produced 216 offspring from mating of heterozygous × heterozygous, 262 offspring of heterozygous black × non-black, 134 offspring of non-black segregates *inter se*, 90 offspring of non-black segregates × pure reds, and 274 pure Durocs. WARWICK explains that the ratios of black to non-black agree very closely with a monogenic hypothesis. Eight black individuals occurred in the progeny of non-black × non-black matings. These can be accounted for by the action of multiple factors which determine the amount of black when the factor for black is present. It is therefore possible for a pig to be free from black but to carry the dominant factor for black.

KRONACHER has an interesting note to the effect that linkage occurs between red and half-colour on the one hand, and black spots and some modifying pattern factor on the other. In this connection it is interesting to draw attention to the Kentucky Red Berkshire which, according to RUSSELL (1922), is indistinguishable from the ordinary Berkshire except in colour. Professor A. L. ANDERSON who has kindly made enquiries concerning these pigs, informs us that this breed is still in existence. It appears to trace directly to the „Red Hog"of the Southern States and is reported to be free from admixture of blood of other breeds. Professor ANDERSON states that in 1930, at the Missouri State Fair, he saw a Berkshire sow that was quite red. That such a combination definitely does occur strongly supports the possibility of linkage between the pattern factor and colour.

Similarly, the possibility of fairly close linkage between „half colour" and red is not to be lightly dismissed. Crossing-over has produced the „half blacks" described by KRONACHER (1924). He mentions the occurrence of half-black pattern in Improved Land-schwein pigs that resulted from the mating of an Improved Land-schwein sow and boar. This is ascribed to the influence of two Bavarian Landschwein sows used several generations previously. On examination, it was found that the Landschwein sow had a Half-red Bavarian as a maternal grand-dam, and the maternal grand-sire of the boar possessed a great number of dark blue spots on its back. The reason given for the occurrence of the half-black is that in the fertilisation of the ova of the sow there occurred an association of factors which combined the factor for black with the characteristic pattern factor of the Bavarian Landschwein.

With regard to the white spotting and white saddle sometimes found in pure-bred Duroc-Jerseys, NORDBY (1934) considers that the white extremities and white saddle must be regarded as two distinct patterns. Further, the genetic expression of either of these patterns cannot be explained on the basis of a single set of factors. The white involved in each case entirely replaces the red. NORDBY is of the opinion that white on the extremities in the Duroc-Jersey is inherited in a similar way to the „six white points" of Poland-Chinas and Berkshires, and the saddle pattern in Durocs is probably inherited in a similar manner to that of the Hampshire. He cites WENTWORTH and LUSH (1923), who found that red pigment depends on the complementary action of two factors, one for sandy and one which intensifies sandy to red, and that the absence of both of these factors produces white. NORDBY concludes his discussion by stating:

„In all probability the explanation for the variations in amount, pattern and location of white will have to be referred both to non-genetic influences and to complex modifiers as they affect the restriction for red." „The inheritance of white-spotting is apparently not due to a simple recessive gene, unless there are factors that determine location, which are inherited independently of the modifiers that restrict the red in such a way as to interfere with the normal ratio."

Despite the intensive experimentation which has been carried out in Germany, there is still room for further enquiry and we would draw the attention of those in charge of experimental herds to a piece of research which can be carried out simultaneously with one of economic importance: the mating of Tamworth or Duroc-Jersey to Berkshire or Poland-China carried into an F_4 would not only provide some extremely useful information as to colour, but would be of assistance in an analysis of the inheritance of economic qualities such as the bacon type, carcase percentage and economy of live weight gain.

KRONACHER has written — „There are great difficulties in experimentation with domestic animals and especially pigs. These difficulties consist partly in that one can hardly find animals that would be homozygous for some doubtful character because in the course of time numerous breeds have been intercrossed and still carry in their germ plasm hereditary factors received from the various breeds."

In regard to its colour, it is the cryptomeric heterozygosity of the pig that is making analysis so difficult, despite the prolificacy of the species. At the moment we must refrain from synthesis. The hope of the writers is that this summary will be of assistance to those who make the further much-required analysis.

6. Colour in relation to other Qualities

There are various reports concerning the association of colour with productivity and hardiness. KRONACHER (1924) states that animals of pure colours, white and black, particularly the latter, are less hardy than those showing red or a combination of two pigments. There are many other similar observations; many of them are contradictory and none are based on sufficient evidence. DARWIN (1868)

says „In the Western countries of England the prejudice ag   ɹ
white pig is nearly as strong as against a black one in Yorkshire'. He
further noted that all the pigs in certain parts of Virginia were black;

.ˈanimals were fed on the roots of the *Lachnanthes tinctoria* which
coloured their bones pink, and, excepting in the case of the black
v .ieties, caused the hoofs to drop off.

.n Great Britain the curers have expressed a definite preference for
white-skinned pigs. This is apparently for two reasons: In the fⁱ
place, because it is claimed that black-skinned carcases are up   .ly,
and secondly, because the black skin is associated, though not
invariably, with „seedy cut" (cf. p. 65).

Professor LUSH informs us that the general experience of breeders
in the United States has been that pigs with white skin are more
susceptible to sunburn and blistering than pigs with coloured
skin. This trouble is not important except in regions where, at times,
there is heavy dew or rain and the pigs in a thoroughly wet condition
are exposed to the rays of the sun. The affection is more extreme in the
southern than the northern states, and has been one of the reasons
why the white breeds have never become numerous in the south.
For this reason the black pig is definitely preferred in many parts of
New Zealand. Against this we may put the fact that of the British
pigs exported to the Northern Territories of the Gold Coast, West
Africa, the white breeds appear to suffer remarkably little from the
sun. Perhaps the solution to the problem lies in the rays of the sun.
Whereas in West Africa it is practically essential that the European
human should wear a sun helmet, in British Guiana, in the same
latitude but on the opposite side of the Atlantic, sun helmets are not
regarded as a *sine qua non*. If this be correct, it is another illustration
of the intimate connection between heredity and environment.

III. HAIR AND SKIN

1. *Hair, Skin and Sebaceous Glands*

TEODOREANU (1931) who made a microscopic study of the hair and
skin of several breeds, finds that, compared with any other breed, the
Mangaliţa has a greater ratio of down hairs to bristles. In the pig
there are two types of sebaceous glands; one is rudimentary and the

ь.   аᵣe located at the base of the legs, tail, down hair and shoulder, wł   the other is fully developed, the glands having five or six lobes. The sweat glands vary with the breed. In the Mangaliţa and Berkshire the body of the gland is built of longish tubules, while in the ì: ʳ· colnshire and Cornwall the tubules are shorter. The same writeᵣ (1930, 1931) has described at some length the structure of the bris   s in different breeds, and in a later paper (1932) discusses the relaı   e ⁻¹iness of the hair of various breeds. He finds the curliness of the Ma   'ita to behave as a recessive in crosses with the Berkshire ıype of coat, and·as an intermediate in crosses with the Cornwall. Density versus absence of hair (Lincoln and Cornwall) also shows an intermediate type of transmission. RHOAD (1934) reports that the curly-haired condition (which he calls „woolly hair") in the native Canasstrao breed of Brazil is clearly inherited as a monogenic Mendelian dominant and in outcrosses no trace of an intermediate condition is found. During the course of an investigation on the F_1 oı Mangaliţa × Large White and × Middle White crosses, CONSTANTINEŚCU (1933) observes that the coat of these crosses is composed of only one type of fibre, whereas the pure Mangaliţa has two types. Curling, although present as in the pure Mangaliţa, appears at a later date. The dark pigmentation of the snout, hoof and skin of the Mangaliţa seems to be recessive.

HÖFLIGER (1931) carried out a detailed study of the hair and skin of the wild and domesticated pıg. He found that the more highly improved the pig, the more it differs from the wild as regards hair and skin structure. The wild type appears to exhibit a greater range of variation in the length, thickness, and pigmentation of the bristles, which are coarser, longer and thicker than those of the improved pig. The stiff bristles are straight, but the soft ones are often slightly twisted. In the wild type, there is also a greater development of the sebaceous glands, and the muscles of the follicles (arrectores pili) are larger; the number of sweat glands is approximately the same but they are more scattered. Domestication appears to reduce the degree of pigmentation, as in the wild type the hair, skin, and hoofs are more deeply pigmented.

2. *Rose*

The term „rose" is applied to the swirl of hair sometimes seen on the posterior end of the back above the loin. CRAFT (1924—26) states that a mating of a sow with a pronounced swirl and a normal boar produced eight swirled boars and no swirled sows. Among 20 F_2 offspring, only two swirled boars were observed. Later (1927—30), the evidence obtained from 268 pigs indicated that the hair swirl is hereditary and is transmitted through both the sire and the dam.

At the Idaho Experiment Station the occurrence of whorls received the attention of NORDBY. In the 100 cases observed by him (1932*a*), affected areas appeared to be limited to the neck coupling at the top of the shoulder, the loin, and the rump region. He analysed data from 27 litters with a total of 253 pigs, 194 of which were free from whorls and 59 had whorls. Further, 30% of the affected pigs were found in 22 litters from sows who either were themselves affected or were known to be carriers. Again, when both parents were affected, 46% of the progeny were also affected. These results are explained by the action of two pairs of factors, probably dominant, W and W', both of which must be present in either a duplex or a simplex state for the character to be manifested. In conclusion, the author states that whorls transmitted by parent to offspring are usually similarly placed on the offspring.

M'PHEE (1932) has obtained whorls by means of inbreeding even when this character was not known to exist in the foundation stock.

IV. PHYSIOLOGICAL CHARACTERS

1. *Blood*

A very limited number of investigations appears to have been undertaken in connection with the physiology of the pig, and still more limited is the work along those lines bearing directly or indirectly upon the genetical aspect. However, brief mention will be made of those papers which were available to the writers.

There are very few references to blood groups in the pig. FISHBEIN (1913) and WESZESKY (1920), (quoted by HERLYN, 1928), identified the presence of groups, which, however, they were unable to classify systematically. SZYMANOWSKI, STETKIEWICZ and WACHLER (1926) examined several hundreds of blood samples and found there were three groups as follows: I AO; II O anti-A; III OO. BERCZT (quoted by DÖHRMANN, 1930) confirms this classification.

DÖHRMANN (1930) failed in an attempt to establish a relationship between breed and the distribution of blood groups. This failure he attributes to the mixture of blood in the course of the development of the individual breeds. His results are as follows:

Breed	Group I		Group II		Group III	
	No.	%	No.	%	No.	%
Mangaliţa	104	38.2	82	30.1	86	31.6
Yorkshire	70	36.6	57	29.8	64	33.5
Berkshire	10	38.5	8	30.8	8	30.9
Improved German Landschwein . . .	9	36.0	8	32.0	8	32.0

SCHERMER (1929, 1930) obtained three groups: one possessing an agglutinogen; one possessing an agglutinin and the third possessing neither. This is represented as follows: I Ao; II Oα; III Oo. (cf. SZY-MANOWSKI, STETKIEWICZ and WACHLER (1926) above). SCHERMER states that the reactions are remarkably definite in the pig compared with other animals.

In an investigation of 628 pigs, SCHOTT (1931) was able to place 592 of them into the three types of blood groups, Ao, Oα, and Oo respectively; the remaining 36 animals could not be grouped. SCHOTT suggests that there may exist a correlation between blood groups on the one hand and fertility and weight of pigling on the other. Further, he states that Mendelian laws appear to apply to the blood groups of the pig.

SCHERMER and KAEMPFFER (1932) carried out experiments to ascertain the mode of inheritance of blood groups in the pig, using the Hannover Improved Landschwein breed. Their results are shown in the following table:

| Parents | No. of Families | Group of Progeny | | | | | | No. of Offspring |
| | | Ao | | Oo | | Oα | | |
		No.	Percentage	No.	Percentage	No.	Percentage	
Ao × Ao	16	56	61.8	18	20.2	16	18	89
Ao × Oα	71	134	52.1	30	11.7	93	36.2	257
Ao × Oo	11	21	65.7	5	15.6	6	18.7	32
Oα × Oα	24	—	—	5	5.6	84	94.4	89
Oα × Oo	13	—	—	12	27.3	32	72.7	44
Total .	135	211	41.1	70	13.7	231	45.2	511

SCHERMER and KAEMPFFER deduce from these results that the factor governing the presence of the agglutinogen A (Group I Ao) is dominant over its absence (Groups II and III Oα and Oo). Similarly, the factor for the presence of the agglutinin α (Group II Oα) is dominant over its absence (Groups I and III Ao and Oo) but it is hypostatic to the factor for agglutinogen A and therefore α-bearing offspring may result from matings of Ao parents. They state that the genes for A and α are not allelomorphs and show no indication of linkage, and conclude that blood groups depend for their inheritance upon the action of two independent pairs of factors $K_A K_a$ and $S_\alpha S_o$. Their experiments are concerned with the three blood groups already described, but in the course of this work they claim to have obtained evidence of the existence of a fourth group, so that the four groups present in man are also present in the pig.

2. Metabolism

A paper of interest is that of DEIGHTON (1929) who carried out metabolic studies on a Berkshire and a Midddle White pig from weaning until maturity. Evidence of an interchange of thermic and dynamic energy (which is partly due to increased activity of the animals when on a sparse diet, and partly to the preferential demand for blood oxygen), has led him to believe that it is almost impossible to calculate an accurate maintenance ration. The metabolic differences are thought by the author to be characteristic of the breed rather than the individual.

3. *Thyroid*

HERMANN (1932), investigating the thyroid of the domesticated and the wild pig, found that the glands show great similarity but with certain exceptions. The rhythm of activity differs in that the range is much wider in the wild pig, while in the domesticated pig thyroid weight is twice as great and the iodine content is considerably lower. He also observed in the domesticated pig that the thyroids of castrates show a tendency towards a resting stage, whilst the normal animals often exhibit disharmonious structure. This evidence should be considered in connection with the problem of the hairless pig *(vide infra)*. Reasoning from the hairless mouse and rat, it is unlikely that the thyroid of the genetically hairless pig differs appreciably from that of the normal pig. It is possible, however, that the definitely lower iodine content of the domesticated pig is contributory to the environmental type of hairlessness. This work may also have a bearing upon the problem of resistance to disease (cf. PALMER 1917) p. 39).

V. DISEASE RESISTANCE

The attempts to produce a strain of pigs highly resistant to disease have not yet met with appreciable success. Investigations of this type involve considerable expense and are therefore limited in number.

The disease which has received the greatest amount of attention is swine fever, as it is known in all English speaking parts of the world, with the exception of America, where distinction is made between the lung and intestinal types of the disease — the former being termed hog cholera and the latter swine plague [1]).

---

[1]) Since a similar distinction is made in France and Germany, the following table of equivalent terms may be useful for reference:

Amongst practical breeders there has for long existed a belief that certain strains, or blood lines, are more resistant to the ravages of swine fever. These beliefs appear to have been first crystallized by MELVIN and SCHROEDER (1906) who during ten years carried out tests which showed that susceptibility to hog cholera varied from an extremely high grade to absolute immunity. Since then, a painstaking investigation of this problem has been made at the Iowa Agricultural Experiment Station by LAMBERT, MURRAY and SHEARER (1928). In this investigation, the foundation material consisted of 1 boar and 9 sows which had been originally bought by a firm of manufacturers of serum and which proved to be refractory to cholera; these animals, therefore, were thought to be naturally immune. The test for immunity consisted of two injections of potent hog cholera virus. During the experimental period of 1924—28, 182 pigs were tested by injection; 11.5 per cent of these recovered from the effects of the dose, but all showed typical symptoms of hog cholera. From these results it is concluded that simple selection alone is not directly effective in increasing resistance. From previous analyses of the data, it was suggested that resistance, if hereditary, must be a complex recessive trait. Four offspring from twelve boars and sows naturally immune to swine fever were tested and only one was found to be resistant.

A similar investigation was carried out at the Illinois Agricultural Experiment Station by ROBERTS and RICE (1924), and by ROBERTS (1925—26). In this experiment also, a very limited degree of resistance was obtained. In 1927, ROBERTS and CARROLL subjected 18 pigs from sows and boars previously selected as resistant to hog cholera, to the hog cholera virus. The test consisted of a subcutaneous

Great Britain	U. S. A.	France	Germany
Swine fever: Intestinal form	Hog cholera	Cholera suum: Peste du porc	Schweinepest
Pneumonic form	Swine plague	Septicémie; Pneumonie contagieuse du porc	Schweineseuche

injection of the virus, a dose by mouth, and contact with a pig which was a virulent case of cholera. All the pigs succumbed.

Another investigation was carried out at Illinois by GRAHAM *et alii* (1927). In this experiment, suckling pigs were given serum and virus in an attempt to immunise them against later attacks of swine fever. The studies failed to show the optimum age for treatment but the younger the pigs, the less the expense of treatment, as less serum is required. Results indicated that there must be other factors beside age which lowered the immunity established by injecting very young pigs with the serum and virus, for in some instances immunity was retained until marketing age was reached, whilst in others this was not the case.

MCARTHUR (1918) states that immunity of sows (produced by the DORSET MILES method) is transmitted to their young during the time they are suckling. This is not in agreement with the work of MINKLER (1916) who found that piglings became infected while suckling their dams which had been previously immunised by the simultaneous injections of serum and virus.

OSSENT (1932) proposed to evolve a new breed which would combine all the good qualities of the improved breeds, e.g. high fertility, rapid growth and fattening capacity, with those of the wild pig, e.g. hardiness and especially resistance to swine plague. For his material, he made various matings of wild boars with the Bavarian Landschwein, Hannover-Braunschweig sows and their F_1. He also used a Berkshire boar. Nearly all the offspring from the matings of wild-coloured with white pigs were either completely or almost completely white, with spots of a light wild and light grey, the skin under the spots being pigmented. The pigs were kept under the worst possible conditions in sties infested with swine plague. All the white pigs died, whilst the wild-coloured ones survived. After rigid selection the remainder were mated to either a wild or a Berkshire boar. This method of breeding was continued until finally mortality and susceptibility to disease were reduced to a minimum but litter size and weight were far below the average for improved breeds. These experiments have been carried out for nearly ten years and approximately 95 per cent of the wild-coloured progeny can be reared, whilst the white animals suffer heavily from swine plague.

NACHTSHEIM (1933) expresses his scepticism of OSSENT's results.

OSSENT (1933), however, points out that it is only the wild-coloured offspring which exhibit this immunity. When considering these results, the important question arises whether these wild-coloured apparently immune animals would retain their immunity if subjected to other conditions in which a different strain of virus was prevalent. From the author's description of the environment it would appear that all generations were reared in the same sties.

Of later work, mention must be made of that of UHLENHUT, MIESSNER and GEIGNER (1933). They tested eight experimental animals belonging to Professor BAUR, of Müncheberg, who was attempting to breed pigs which would be resistant to swine fever. The animals were cross-breds and contained Mangalița, Güstin Pasture, Berkshire and Improved Landschwein blood. These pigs succumbed to the disease both when kept with infected pigs and when subjected to injection of the virus. Further tests were made with an F_1 litter from a pure-bred Mangalița boar × a Güstin Pasture-Berkshire-Improved Landschwein sow belonging to the Kaiser-Wilhelm Institut für Züchtungsforschung. When only a weak virus was used, these pigs suffered less heavily than the controls. When tested by injection with strong virus, however, the animals rapidly developed acute swine fever. Animals which were simultaneously injected with virus and vaccine proved to be highly resistant to swine fever.

The varied results obtained by different investigators have shown the extreme difficulty that besets an experiment of this nature, largely due to the difficulty of standardising the degree of infection to which the pigs must be subjected. A certain degree of resistance seems to be inherited, the mode of inheritance depending on multiple factors acting in a recessive manner. It would appear unlikely that an absolute immunity to the disease exists in nature.

PALMER (1917) found that extirpations of the thyroid gland did not induce cretinism and did not retard growth; except for a markedly lower resistance to infection, the pigs behaved similarly to the controls.

CONSTANTINESCU (1933) states that cross-breds (Middle White × Mangalița) withstand infection better than the Middle White. In two places in Brazil, pigs were found (KUCHENBECKER 1931) which resembled the solid-hoofed pigs, and these were highly resistant to foot-and-mouth disease. CRAFT (1931) states that inbreeding tends to make pigs more susceptible to worms, „influenza", pneumonia and enteritis. Inbred pigs also yield less readily to worm treatment. BECK (1933) found by experience that the Yorkshire had less sickness and greater powers of resistance to disease than the Danish Landrace.

VI. MENTAL TRAITS

The performing pig is not unknown to the older generation and is yet to be found on the continent of Europe.The intelligence displayed by the best of these can hardly be excelled by the ape. How much of this intelligence is due to nature and how much to nurture has not yet been decided, but the SIMPSONS (1911) have published some interesting remarks on two strongly marked mental traits of the species. They state that both the Berkshire and the Yorkshire (Large White) in their skull formation are typical of their Chinese progenitors. The Chinese hog has for so many centuries been kept in sties that it has lost „the filial attributes of the gregarious Suidae that was necessary to its preservation when in nature". The gregarious nature of the Tamworth is well known. „An obstinate Tamworth, wild Arkansas, or a German wild, or their hybrids, may be quickly brought back with the moving herd by an offensive attack of the collies, whilst a Berkshire or a Yorkshire must be brought in by a spreading of jackets and a careful ‚soo-o-o-boy'." The father of these writers rarely failed to detect the otherwise invisible Berkshire taint by the fact that the pigs could scarcely be driven through a hole in the fence. „Yorkshire hybrids may be sorted back as with a sieve in driving a mixed drove of swine over a planked railway crossing."

KRONACHER (1930) describing an F_1 from a wild boar and an Improved Landschwein sow, found the wild instincts highly developed particularly as regards the sense of smell, expert way of foraging for themselves, and rough treatment of sows. Pigs of such a cross are described as capable of passing through the smallest possible opening in order to gain freedom. The wild instincts appear to be associated with the wild colour. Mention may also be made of the use made of pigs in searching for truffles.

The Danish type of pig house containing a dung passage to which the pigs have free access, shows, from the experience of the present writers with one constructed in Edinburgh, that some strains of pigs behave in a far more cleanly manner than others.

The remarks of MANZANO (1934), a well known animal trainer, are of interest. He states that performing pigs have been a feature of old English fairs and circuses for the past hundred years. The pigs which he trained were selected at random from ordinary farm stock. Perfection of performance was reached by kindness and patience, but never force. MANZANO found that pigs, compared to other animals, possess great self assurance and lack of fear of lights, horses, etc.

VII. SEX

As an economic producer of food for human consumption, the pig ranks next below the dairy cow and the laying hen. This is primarily due to the prolificacy of the porcine species. In passing, it might be mentioned that these three animals depend for their economic production upon some aspect of sexual activity. The value of the pig depends on the prolificacy of the sow.

1. *Sterility*

In cattle and horses the connotation of sterility is comparatively precise. In this paper it is intended to use the term in the absolute sense of inability on the part of a sow and a boar to produce any young whatsoever, fertility being the opposite. Prolificacy may be defined as the number of progeny of a given mating, the greater the number of offspring, the higher being the degree of prolificacy. In popular language, a sow producing less than six offspring at one parturition is considered unprolific.

KRONACHER(1924),in an inbreeding experiment, found that closely related individuals will not mate at all. The experience of DE-CHAMBRE (1919) has led him to believe that, as a general rule, the matings of domesticated and wild pigs are fertile. He reports a case of sterility where a wild sow refused to hold to a domesticated boar, but this is more likely to be of nutritional than genetic origin. According to DAMLE (1931) and PATEL (1932), wild pigs have only one breeding season per year.

FUNKQUIST (1919) analysed a strain in which there occurred boars which were unable to perform mating though they produced fertile sperm. From the study of pedigrees it is concluded that this type of impotence is inherited in a recessive sex-linked manner.

2. *Prolificacy*

a) The Influence of Non-Genetic Factors
Age.
Prolificacy is influenced by other factors besides genetic. Such workers as JOHANSSON (1931) and STEWART (1931) state that the genetic aspect is of very secondary importance. As in the case of the milk production of the dairy cow and the prolificacy of the laying hen, age is an important factor in the sow. From the records of 278 farrowings, SINCLAIR and SYROTUCK (1928) found a marked rise in litter size from one to two years of age and a less marked increase thereafter. Advancing age was associated with a greater number of still births and of crushed pigs, and also an increase in the birth and weaning weights of the offspring. Their figures are as follows:

Age in Years	No. of Sows	Average Litter Size
1	101	8.109
2	77	8.701
3	57	9.706
4	33	9.97
5	15	9.53

These figures are in essence borne out, though at somewhat different levels, by other investigators including BARTRAM (1926) (who also states that the variability in litter size tends to increase with age), KEITH (1930), CARMICHAEL and RICE (1920), ELLINGER (1921), MACHENS (1915), STAHL (1930), AXELSSON (1928) and JOHANSSON (1929a, b, 1931). On the other hand, KRALLINGER & SCHOTT (1933) found that the percentage of fertilisation falls with the sequence of matings. Figures quoted are as follows: the percentage of fertility for the 1st—3rd litters = 84.9 per cent., the 4th—6th litters = 64.8 per cent., and the 7th and later litters, 55.6 per cent.

MUMFORD et alii (1924, 1925) report that for a number of years sows were bred at the earliest possible date in order to determine the effect upon the subsequent generations. The 13th generation of early matings farrowed at 10 months 12 days. Up to that period there was no evidence of injury to either race or breed. This is illustrated by the fact that a sow of the 9th generation reached 687 lbs. at the age of 4 years 9 months 12 days and farrowed 17 pigs in her 7th litter.

DASSOGNO (1915) reports that age affects the length of the gestation period. This is confirmed by JOHANSSON (1929c) who, however, found the length of gestation to be only slightly increased with age. ZORN, KRALLINGER & SCHOTT (1933) find that litter size increases until the 6th litter, after which it falls.

KEITH (1930), working on the breeding and farrowing records of the herds at the Illinois Agricultural Experiment Station from 1903 to 1925, which include 935 litters containing 8478 pigs of all the principal American breeds, found that the number of the pigs in the litter increases with the age of the dam up to about 4.5 years, after which there is a gradual decline. He obtained a significant degree of correlation between the size of a litter farrowed at a given age of the dam and the size of a litter farrowed at later ages. A high correlation was found to exist between the size of the second litter and the average size of succeeding litters. There was greater variability in the size of the first litters than in that of the second. M'PHEE (1931) also found a correlation between the size of first litter and that of later ones when $r = + 0.2051 \pm 0.0409$.

Other influences

There is also evidence in favour of litter size being influenced by the season of the year, the interval between gestations, nutrition, and other factors. AXELSSON (1928) is of the opinion that season has no influence upon litter size at birth or upon subsequent growth. PEARL (1918) states that from Poland-China and Duroc-Jersey gestation records, most pigs appear to be born in the spring, with the exception of those in the southernmost states of the U.S.A. In passing, mention must be made of some evidence (largely conflicting) to the effect that the interval between gestations has little influence upon subsequent litter size.

It is interesting to note that DASSOGNO (1915) found the size and weight of litter to have no effect upon the gestation period. This was confirmed by Johansson (1929c) who also found that the season of the year had little influence on the gestation period.

Ossent (1932) records that the length of gestation of wild-coloured pigs was 110—114 days, whilst that of the improved breeds was 120 days.

b) The Influence of Genetic Factors
Relationship Correlations

Prolificacy is inherited probably in a straightforward manner — though perhaps not too simple. The sire and the dam would seem to play an equal part. In the production of a specified litter, the boar, if he be normal and in good condition, may have little influence on the numbers produced by the sow (Krallinger & Schott 1934), but he does influence (probably in every way as much as the sow) the size and type of litters produced by his daughters.

Passing mention must be made of correlations of litter size of one generation with another, i.e. daughter to dam and to grand-dam. Rommel and Phillip (1907) correlated the litter sizes of two successive generations and found a small but appreciable correlation of $r = 0.06 \pm 0.0086$ between the litter size of dam as compared to that of grand-dam; the correlation was higher when the first litters of gilts were compared with the litters of the dams, r being equal to $+ 0.1008 \pm 0.0149$. They state: „We are consequently justified in concluding that litter size is transmitted from mother to daughter". They were unable to determine the part played by the sire. Wentworth & Aubel (1916) are of the same opinion concerning the influence of the dam on the size of her daughter's litter. Smith (1930 —31) concludes that the boar does not influence the litter size of the sow to which he is mated but that he does affect the size of his daughters' litters. Zorn, Krallinger & Schott (1933) state that the sow plays the decisive role in determining litter size.

Morris and Johnson (1932) obtained a very low degree of correlation between the size of the litter in which the dam was born and the size of the litter produced by her. M'Phee (1931) correlated the size of the litter in which the dam was farrowed with that in which the boar was farrowed and found $r = + 0.3471 \pm 0.0244$. Furthermore, in correlating size of litter produced with size of litters in which sire and dam were farrowed, $r = + 0.2964 \pm 0.0254$, and $+ 0.0702$

± 0.2725, respectively. (The writer adds that this unexpected result, of no correlation between size of litter in which the dam was farrowed and her litter, was checked.)

The Litter in Utero.

A line of investigation which might throw much light upon this question of prolificacy is that dealing with the litter in utero. For obvious reasons only a very few experiments of this type have been undertaken. HAMMOND (1921) examined the reproductive organs of 22 pregnant sows. There were 396 corpora lutea (or an average of 18 per sow) but only 267 of these (or 12.1 per sow) had developed into normal foetuses when the sows were killed. There were 49 atrophic foetuses (or 2.2 per sow), thus leaving 80 ova (or 3.7 per sow) unaccounted for. The author considers that these were either unfertilised or had perished at a very early stage and were absorbed. He suggests that fertility in pigs is mainly influenced by those factors which control the number of eggs which develop. CORNER (1921), from records including 4480 corpora lutea, calculated a discrepancy of 23.3 per cent. WARWICK (1927) examined the uteri of 448 sows which where unfortunately of unknown breeding. The result of this examination showed that 3.68 per cent. of the foetuses observed were in various stages of degeneration, with more or less resorption. Overcrowding could not completely explain this condition, as degenerating embryos were found where there was no evidence of overcrowding. Genetic causes were suggested.

CREW (1925) observed a higher male than female prenatal mortality.

DAVIDSON's (1930) work is also of interest. He fed calcium and protein deficient rations to two groups of 6 gilts in each group. He concludes that partial foetal atrophy is not due to protein deficiency and that calcium deficiency may be a contributory but not a major factor.

c) The Influence of Breeds and Breeding Methods upon Litter Size and Weight.

The evidence of practically all the recent investigators in Europe and America goes to show that while there are possibly slight variations between breeds, the average litter size at birth of the better known breeds of pigs is between 8 and 11.

BITTING (1897), working with herd book data, found that various breeds of pigs in the United States had distinctive litter sizes; that in two breeds the average number per litter had increased since the foundation of the herd books; and that in a third this figure had decreased. However, his figures are by no means conclusive.

ROMMEL (1906), also working from the herd book records of two American breeds, the Poland-China and the Duroc-Jersey, found that in the former breed the average litter size during the twenty preceding years had increased by nearly 0.5, while there was no change as regards the latter. The former breed had an average litter size of 7.52 and the latter 9.26. The numbers on which these figures are based seem adequate. These results have been quoted on both sides of the Atlantic as evidence that litter size is not inherited, though the author does not state this in his work.

SURFACE (1909), who was an expert statistician, examined the figures of ROMMEL and found that undoubtedly real breed differences do exist. He gives the following constants for variation of fecundity in the Poland-China and Duroc-Jersey breeds, as derived from ROMMEL's figures for 1902:

Constant	Poland-China	Duroc-Jersey
Mean	7.435 ± 0.010	9.337 ± 0.021
Standard deviation	2.038 ± 0.013	2.427 ± 0.016
Coefficient of variation . . .	27.411 ± 0.172	25.997 ± 0.169

GEORGE (1912) confirms these findings.

Records at the Institute of Animal Genetics, Edinburgh, show the existence of prolific strains of leading British breeds. Whenever we have heard of remarkable prolificacy in a pedigree sow, we have attempted to obtain as much information as possible about her. In the great majority of cases, these sows come from strains which are also remarkable for prolificacy, their daughters and many of their granddaughters having been prolific above the average. We have studied cases where there is a lack of prolificacy, a state which clearly runs in certain strains where the average litter of a sow is about four.

WENTWORTH and AUBEL (1916) made an extensive investigation of this subject. The frequency curves for the 3,540 litters studied lead them to believe that there are at least three centres of deviation in swine fertility. These possibly correspond to the genetic factors involved in its inheritance. It has been claimed by them that small litter size is dominant to large. The data on which this observation is founded were drawn from crosses involving wild pigs. Few races of feral swine are prolific. SIMPSON (1912) crossed a German wild boar with a Tamworth sow. The German wild pig has an average litter size of four and the particular strain of Tamworths averaged about eleven. The F_1 litter consisted of 9 offspring. Three of the cross-bred daughters were mated, one to her litter brother, by whom there were four piglings; the other two, to unrelated Tamworth boars, with results of litters of five and six. The daughter which produced the litter of six was then mated to a wild boar and farrowed seven pigs. One of the litter of six (crossbred × Tamworth) was mated to an unrelated Tamworth and gave birth to twelve. Similar evidence was obtained from analogous experiments by WENTWORTH and LUSH (1923) and CULBERTSON and EVVARD (1925).

CHRISTENSEN, THOMPSON and JORGENSON (1926*b*), in breeding records dating from 1909 and including the progeny of 393 sows, found considerable breed variation in both litter size and mortality. This can be seen in the following table:

	Average Litter Size	Per cent. reared
Yorkshire	11.7	74%
Duroc-Jersey	10.7	67%
Chester White	9.6	73%
Berkshire	8.7	69%
Poland-China	8.2	65%

69% of the mortality occurred during the first week of life.

STAHL (1930) and JOHANSSON (1931) both give figures for postnatal mortality. JOHANSSON's figures, which are based on 1,671 litters from forty-nine herds of Swedish Large White pigs, are as follows:

Age of sows (expressed by the consecutive number of litters) in relation to size of litter at birth, weight of living pigs at three weeks of age, and death rate in litters during the first three weeks after birth. (From JOHANSSON)

Litter	Number of Litters	Pigs born alive per litter	Living pigs per litter	Death-rate %	Three weeks after birth			
					Average weight			
					Litter		Pigs in the litter	
1	2	3	4	5	6		7	
					Kgs.	lbs.	Kgs.	lbs.
I	287	9.43	8.17	13.4	40.5	89.30	4.96	10.94
II	321	10.59	8.98	15.2	46.4	102.31	5.16	11.38
III	307	10.99	8.98	18.3	46.0	101.43	5.13	11.31
IV	259	11.05	8.79	20.5	45.5	100.33	5.18	11.42
V	175	11.41	9.05	20.7	46.5	102.53	5.13	11.31
VI	126	11.06	8.60	22.2	43.5	95.92	5.06	11.16
VII	83	11.11	8.52	23.3	42.9	94.59	5.04	11.11
VIII	54	10.78	7.72	28.4	39.8	87.76	5.15	11.36
IX—XV	59	10.24	7.85	23.3	39.9	87.98	5.09	11.32
Total . .	1,671	10.68	8.68	18.7	44.3	97.79 [1]	5.11	11.25 [1]

STAHL (1930) confirms JOHANSSON in stating that first litters have a lower mortality but that this rises with the seventh and eighth farrowing. RACZ (1931) is of the same opinion.

In recent years, as reported in the Danish Foreign Office Journal (1930), considerable improvement has been made in Denmark in the grade of fertility and in the production of litters of greater vitality. The average number of pigs per litter for all the Landrace breeding centres increased in the period 1921—29 from 10.6 to 11.3, while the vitality rose from 75% to 77%. Since these stud records now go back over 20 generations, a high degree of accuracy is possible in the selection of breeding stock.

[1] 1 Kg. = 2.205 lbs.

LÜTHGE (1933) investigated the question of uniformity in litters. For his material he had 7 families with 86 sows of the Improved Land-schwein and 5 families with 21 sows of the Berkshire breed. The uniformity of the progeny of each sow was studied. Great variability was observed as regards individual litters; some sows produced litters which showed marked uniformity, whilst others were the reverse. Mortality was found to increase with the fall in weight below the average, while it was low among the heavier classes. In conclusion LÜTHGE emphasizes the importance of eliminating, for breeding purposes, sows which produce a high proportion of small piglings.

WILD (1927) could find no significant differences between the Edelschwein and Landschwein regarding the litter size, mortality and number of pigs reared to weaning. He maintains that birth weight is no indication of subsequent gain in weight.

AXELSSON (1928), from 2222 litters of the Swedish Large White and 395 litters of Swedish Landrace, calculated the following average litter sizes:

	Average number at birth	Average number at 3 weeks
Large White	10.1638 ± 0.0587	8.0968 ± 0.0491
Swedish Landrace . . .	9.9367 ± 0.1286	8.3063 ± 0.1089

The correlation of litter size at birth and at 3 weeks was $r = + 0.6765 \pm 0.0115$ and 0.7085 ± 0.0251 respectively.

Several other workers have investigated the problem of breed or type in connection with the size and vigour of the litter. Amongst these may be mentioned the Iowa Station (1925), where a small difference was found in the farrowing weights of big, medium, and small type pigs, the Kentucky Station (1922), HICKS (1922), and TINLINE (1922).

RACZ (1931) states that by means of selection, the litter size of the Mangaliţa breed has increased from 5.9 to 6.3 from 1928 to 1930. By crossing Mangaliţa with Berkshire or Cornwall, prolificacy was in-creased by 20 per cent. Similarly, MORRIS and JOHNSON (1932) demonstrated, by the results of their analysis of 1,035 litters taken at random from Poland-China records, the improvement which can be achieved by means of selection. In this instance, however, selection was a more gradual process, the average litter size being increased by 1 during the period 1900—1920.

MOHLER (1933) described an experiment carried out with different types of Poland-China pigs. Small, intermediate, and large type pigs were used. „In the spring of 1932 five small-type sows farrowed litters averaging 4.2 pigs and weaned an average of 2.0 pigs. Six intermediate-type sows farrowed an average of 6.7 pigs and weaned 5.8 pigs. Five large-type sows farrowed an average of 6.8 pigs and weaned 5.4 pigs".

The following table of OSSENT (1932) is of interest in that it also shows the great improvement in both size and weight of litter that can be achieved by selection in animals of mixed breeding.

	Litter Size	Weight		
		Birth Kgs.	4 weeks Kgs.	10 weeks Kgs.
1930	6.1	0.84	5.7	15.0
1931	7.4	0.975	6.3	17.20
1932	8.0	1.125	7.1	—

Similarly, the statement of WILEY (1926) is also striking evidence of the effect of selection upon litter size. By means of selecting sows for breeding from litters of 10 or more, the average litter size was raised from 5.0 to 7.11 during the years 1922—26.

SMITH (1930—31) gives the ideal size of a litter as from 10 to 12, but in view of the results of other workers, it would appear unwise to fix any definite figure owing to the interaction of other factors. In contradiction to JOHANSSON, KEITH (1930) states that size of litter is a valuable criterion in selection for fertility.

VITZTHUM VON ECKSTAEDT (1928) practised inbreeding and found its effect upon prolificacy to vary with the individual: in some cases prolificacy was improved and in others it was lowered. Data collected from 1200 sows and 400 boars showed great variation in the fertility of different females.

SCHMIDT, LAUPRECHT, and VOGEL (1926), from data drawn from 163 litters of Improved Landschwein, found the birth weight of individuals in litters of 9 or more to be 1.23 kgs., as compared with 1.37 kgs. in litters of 8 or less. JOHANSSON's (1931) figures also demonstrate that the smaller the litter, the greater the average weight of the pigs. MURRAY (1934) obtained the following figures for the birth weights of pigs in litters of different sizes:

BIRTH WEIGHTS OF PIGS

Litter size	6—8		9—11		12—14		15—17	
Sex	M.	F.	M.	F.	M.	F.	M.	F.
No. of pigs . . .	38	26	108	103	79	64	22	20
Average weight, lb.	3.5	3.17	3.08	2.75	2.69	2.67	2.34	2.21

AXELSSON (1928) calculated the correlation between the number of pigs at birth and the total weight of the litter and obtained $r = 0.815 \pm 0.0322$. The larger the average fertility, the stronger was the negative correlation between litter size and weight of individual animals at 3 weeks. Furthermore, the larger the litter, the greater the proportion of undersized pigs. The way in which the negative correlation decreases as growth proceeds can be seen in the table below:

Age of pigs (in weeks)	Correlation coefficient between number of pigs born and the average weight of pigs in litter r
Birth	—0.406 $\pm$ 0.0583
1 week	—0.433 $\pm$ 0.0556
2 weeks	—0.441 $\pm$ 0.0548
3 „	—0.406 $\pm$ 0.0582
4 „	—0.313 $\pm$ 0.0673
5 „	—0.307 $\pm$ 0.0680
6 „	—0.235 $\pm$ 0.0750

MURRAY (1923) studied the influence of size of litter on total litter weight at 8 weeks and found that the litter weight increases with increase in litter size up to 12, after which the weight decreases. His actual figures are given in the following table:

Influence of Litter Size on Litter Weight at 8 weeks. (From MURRAY)

Litter size	6	7	8	9	10	11	13	12	14	15	16	17
No. of litters . .	3	3	3	5	7	8	7	7	2	1	1	1
No. of pigs weaned	14	20	20	34	56	59	63	63	18	6	9	12
No. of pigs weaned per litter . . .	4.7	6.7	6.7	6.8	8.0	7.4	9.0	9.0	9.0	6.0	9.0	12.0
Total litter weight at 8 weeks, lb.	173	227	240	248	256	225	309	264	243	179	215	336

JOHANSSON's figures (1931), which are based on a larger number of litters, show that up to three weeks of age the death rate steadily increases with the size of the litter, while STAHL (1930) was unable to obtain any degree of correlation between litter size and percentage survival up to the time of weaning.

MOHLER (1933) gives the results of an interesting experiment in which the average litter sizes of gilts from gilts, gilts from old sows, and old sows are compared. The actual figures are:

	Average litter size	Percentage of pigs weaned
Gilts from gilts	7.7 pigs	72.7
Gilts from old sows	8.9 ,,	76.5
Old sows	10.2 ,,	.61.9

d) N u m b e r o f M a m m a e

Before selecting for increase in litter size in breeds such as the Large White, which are already prolific, JOHANSSON (1931) stresses the importance of selecting for an increase in the number of functioning mammae. It is useless to obtain a high degree of prolificacy in a sow which is unable to bring a large proportion of her litter to the weaning stage.

There seems to be a certain amount of evidence in favour of the existence of a correlation between the number of mammae and litter size. Apart from the genetic aspect, there would appear to be no little probability of such a relationship, as obviously the chances are much in favour of the sow with a well developed mammary system rearing a large litter with success.

KONOPIŃSKI (1932) calculated the coefficient of correlation between number of teats and the litter size of the sow. From data which included 1068 unselected pigs of various types, he obtained a correlation of r = + 0.20 ± 0.019, and with 972 selected pigs of the Edelschwein type, r = + 0.1324 ± 0.0246. Lastly, with data from East Poland and 725 native or Yorkshire pigs in Denmark, r = — 0.024 ± 0.021. Unfortunately the results are not entirely consistent, which, however, is characteristic of genetic investigations of the pig, for correlations and associations of genetic factors found to exist in one breed or type of pig do not hold good in the case of other breeds or types. The reaction of this fact on the results of nutrition experiments deserves some consideration.

TEODOREANU (1932) is of the opinion that the size of litter and the number of teats show an intermediate type of transmission.

The difficulty of tracing the inheritance of a sexual character, such as prolificacy, lies in the fact that the phenotype of one sex in respect of that particular character cannot be assessed. The question of the inheritance of nipples, however, would appear to form a welcome exception. To WENTWORTH (1912) goes the credit of the first investigation of this problem *(vide infra)*. He was followed by PARKER and BULLARD (1913) who, in a population of 5790 foetuses, found the number of mammae to vary from 9 to 18, with a mean of 12.4 and a standard deviation of 0.6906±0.0060, while for females the respective values were 8—18, 11.9, and 0.7905 ± 0.0069. On the left side, the number of nipples varied from 4 to 9, with a mean of 6.1, while on the right, it varied from 4 to 10, with a mean of 6.1. In 3559 cases, the arrangement of the nipples was regular and in 2411 irregular. They could find no relationship between prolificacy and the number of mammae.

WENTWORTH (1912, 1913) and WENTWORTH and LUSH (1923) give evidence which clearly points to genetic factors. WENTWORTH found 130 pigs to be symmetrical while 68 were asymmetrical. No connection could be found between the dams and their offspring as regards asymmetry. In all but three cases, the extra mammae appeared on the left side. The writer postulated at least two pairs of factors.

WENTWORTH (1914) also published an interesting paper in which the data dealing with the presence and absence of rudimentary teats in both males and females were discussed. Boars possessing rudimentaries were mated both to similar and to normal sows, and vice versa. From the results of these matings, he concludes that the mode of inheritance appears to be a combination of both the sex-linked and sex-limited types.The factor (or factors) for rudimentaries is transmitted in a sex-linked manner, but are sex-limited in that they do not become apparent in the female when in the simplex state. If X represents the sex-chromosome and R the factor for rudimentaries, the following types are possible:

Boars	Sows
XRO	XRXR
XrO	XRXr
	XrXr

Only the first type of each sex would have rudimentaries somatically. The following table shows the results of matings of animals which do not manifest rudimentary nipples:

	Males with rudimentaries	Males without	Females with rudimentaries	Females without
XrO—XRXr	19	16	0	17
XrO—XrXr	0	32	0	31

This work was continued by WENTWORTH and LUSH (1923), who obtained results which failed to support sex-linkage in the inheritance of rudimentaries and caused sex-limitation to appear extremely doubtful.

NACHTSHEIM (1925a) has made an extensive investigation of this subject and classified nipples into three types: normal, supernumerary and pseudo-nipples. He mentions that the 2nd and 6th pair of normal nipples are absent in the European wild pig (Sus scrofa), and points out that in the two more primitive German breeds, the Half-red Bavarian and the Hannover-Braunschweig, which may be regarded as being descended from Sus scrofa, these two pairs of nipples are more often absent than in other breeds. This author states that the

2nd and 6th pair of nipples are probably present in the Asiatic pig (*Sus vittatus*) and are definitely present in its domesticated descendant, the Chinese Mask pig. Since the European domesticated breeds are derived to some extent from *Sus vittatus*, he suggests that the factor for the 2nd and 6th pair of nipples originates from *Sus vittatus*.

In the analysis of 1000 offspring of eight boars, NACHTSHEIM found the number (excluding pseudo-nipples) to vary between 10 and 17, with a mean of 12.8. The largest group, consisting of 28.8 per cent. of the whole, fell in the 14-nipples class. The progeny of the separate boars varied significantly. No sex-dimorphism was found to exist as regards the number of normal and supernumerary mammae. The average number of nipples in the male was found to be 12.58 and in the female 12.51. A positive correlation was observed between the right and left side, the average number on the left being somewhat smaller than that on the right, both as regards the whole population and in the progeny of the separate boars.

RACZ (1931) observed that in pure-bred Mangaliţa sows the number of mammae varies from 10 to 12, while in cross-breds from Mangaliţa sows and German Improved boars, the number is increased. The position of the 6th pair is said to vary considerably, and when there are 5 pairs, the 3rd pair (numbering from front to rear) gives more milk than the others. Further, he has found a direct ratio between fecundity and number of teats, sows with 12 teats being 30 % more prolific than those with 10. This increase in prolificacy seems somewhat high and would appear to need further confirmation. HÖFLIGER (1931), in a study of the wild and domesticated types of pig, states that the mammary gland of the wild type has fewer teats which vary less in number than those of the improved type.

3. *Mothering Ability*

As regards economic production, good mothering ability, which includes milking capacity, is probably of even greater importance than prolificacy, though at the same time it must be remembered that neglect to breed for prolificacy may result in smaller litters.

There is little scientific evidence to show that good mothering is inherited, but the fact that it is stressed under the various pig testing schemes, demonstrates that practical experience believes this to be the case. M'PHEE (1930a) states that bad-tempered pigs have arisen as a result of inbreeding. Our own observations in pigs confirm the contention that the ability to nurse the young is inherited. There are breeders in the United Kingdom who believe that bad mothering is frequently associated with the earlier maturing types of pig. It is not difficult to credit a physiological relationship of this nature.

KING (1926) states that mothering ability is hereditary and is transmitted through both the male and female. SMITH (1930—31) is also in support of this theory. HANSSON and BENGTSSON (1926) have found that Yorkshire sows make as good mothers as the Swedish Landrace. GRIMES and SEWALL (1930) consider that the more prolific sows are better mothers and raise a larger percentage of their young. As a result of various experiments, OSSENT (1932) concludes that wild-coloured sows are excellent mothers, careful with their young, docile and generally tolerant towards others. JOHANSSON (1931) emphasizes the importances of selecting for good mothering ability. KULOW (1928) practised line-breeding from two sows and observed definite differences regarding the mothering ability between the two lines, although the fertility and number of offspring from both lines appeared to be about the same. GRANDI (1931) states that Large Black sows proved to be of superior mothering ability to the Middle White, in spite of the fact that the former breed was the more delicate.

4. *Milking Capacity*

As already stated, milk producing ability is closely allied to good mothering, and there is evidence from Germany to show that milk yield is dependent upon genetic factors.

SCHMIDT and LAUPRECHT (1926) have studied milk production in the Landschwein breed. The yield appeared to be higher in sows producing large litters than in those producing small litters; there was also great individual variation. Furthermore, the anterior „quarters" of the mammary gland were more highly productive than the posterior. SCHMIDT, VOGEL and ZIMMERMANN (1929) could find no significant difference between the milk yields of the Improved Landschwein and the Edelschwein.

RICHTER, HEMPEL, OHLIGMACHER and RODEWALD (1928) jointly

and severally carried out a painstaking study of milk production in sows. They found great individual variation in yield and composition of milk. As a measure of productivity, the weight of the sow was used. Heavy milking sows lose a considerable amount of weight; the loss appeared to be great during the first 4 weeks; the weight then was maintained at about the same level and finally rose gradually. RICH-TER, working with the Edelschwein and the Landschwein, weighed 60 sows of each breed and found the loss in weight to be the same: with similar weights at farrowing, the Landschwein subsequently lost 8.7 kgs. and the Edelschwein 9.1 kgs. HEMPEL concludes his observations by stating that (1) milking capacity in sows is inherited, (2) individual variation is to be found both as regards total yield and the shape of the lactation curve, and (3) the amount of milk produced increases with litter size.

OHLIGMACHER confirms the observations of HEMPEL. He also believes to have found indications of hereditary transmission of fat-content. The average albumen content was found to be 6.25 per cent, but by rich protein feeding, it could be raised by as much as 2 per cent. Less variation was found in the albumen content than in the fat content of normal milk.

HEMPEL gives a summary of previous work on milking capacity of sows and the following details may be of interest. OSTERTAG and ZUNTZ (1908) found the amount of milk taken by one pigling at one suckling to be 60—75 gms. According to GOHREN, a Yorkshire sow produced 1.375 Kgs. of milk in 24 hours during the fourth week, whilst CARLYLE found the average daily milk yield of 12 sows (Berkshires, Poland-Chinas and Razor Backs) to be 2.826 Kgs. during the 4th week and 1.746 Kgs. during the 8th week. SCHMIDT estimated that 2 sows during the second half of suckling period produced 4.61 Kgs. and 5.27 Kgs. respectively, while other 3 sows during 8 weeks produced a total of 183.68 Kgs., 208.08 Kgs. and 272.16 Kgs.

DECHAMBRE (1934) found that the milk yields of sows of various breeds, for a period of 84 days and daily, were as follows: Berkshire, 204.27 Kg. and 2.862 Kg.; Poland-China, 194.6 Kg. and 2.204 Kg.; Yorkshire, (daily yield only), from 1.860 Kg. to 2.625 Kg.

RACZ (1931), amongst many interesting observations, concludes that uniformity of litter is very closely correlated with the equality of the yields of milk from each of the teats. Regarding the quantity of milk produced, he estimated the yield from 7 Mangaliţa sows to vary from 119.31 Kgs. to 190.55 Kgs. during a period of 62 to 75 days.

SANDERS (1931) also draws attention to the great individual variations in milk yield. He points out that this variation can be detected by means of pig-recording and the results used to facilitate selection. AXELSSON (1929a) states that the range of variation for milk yield and fat yield is no greater than that for other characters.

5. *Sex Ratio*

WILCKENS (1886) found the sex ratio at birth to be 52.09 per cent. males. ZORN (See KRALLINGER 1930) obtained a ratio of 60 per cent. PARKES (1923a, 1923b, 1926) obtained the figure of 51.15 per cent., while CARMICHAEL and RICE (See McPHEE 1925—26) observed 52.06 per cent. McPHEE (1927) shows the pitfalls in herd-book data which have formed the basis for most of the previous observations. The figures he accepts as reliable are as follows: CARMICHAEL and RICE 51.96 per cent.; SEVERSON 52.3 $\pm$ 0.0056 per cent.; and McPHEE 51.99 $\pm$ 0.0038 per cent. CREW (1925) obtained a primary sex-ratio of 54.55 per cent. and a secondary sex-ratio of 50 per cent., the latter figure being obtained from the records of 1472 newly born piglings. This illustrates the fact that as pregnancy proceeds, the sex-ratio swings from inequality to approximate equality. Prenatal mortality is therefore much greater among males than among females.

PARKES (1925) in his examination of 583 foetuses collected from the uteri of pregnant sows at the Islington (London) Abattoir, found 56.8 $\pm$ 1.38 per cent. of these to be male. The ratio varied with the size of the foetus. Under 100 gms., the percentage of males was 59.1 $\pm$ 1.98, from 101 to 300 gms. it was 57.0 $\pm$ 3.12, while 301 gms. and over gave a sex-ratio of 53.2 $\pm$ 2.45 per cent. The sex-ratio of the foetuses from the two cornua were not found to be significantly different. PARKES estimated the ratio at conception to be approximately 60 males: 40 females. CHRISTENSEN, THOMPSON & JORGENSON (1926b) calculated from their data that the percentage of males to females was 52.3 per cent. SCHMIDT, LAUPRECHT and VOGEL (1926), with records of six groups of pigs of different breeding from 248 observed litters, obtained a sex-ratio of 50.17 per cent., and a smaller proportion of males in litters of more than 12 pigs. This is confirmed

by KRALLINGER (1930) whose work deals with the German Improved
Landschwein and Edelschwein and appears to have been taken
from published records. He obtained 50.57 ± 0.23 males at birth
and found no difference between the two breeds.

KRALLINGER's most interesting finding is a high difference in the
sex-ratio amongst the progeny of different boars and he also considers
it probable that the sow plays a part. Further, KRALLINGER finds a
negative correlation between the sex-ratio and the degree of prolificacy: the larger the litter the smaller the percentage of males. This is
probably true but further confirmation is required. He also finds the
sex-ratio to be somewhat higher in first litters. No seasonal variation
was observed. Later results (1933) indicate that time of mating has
no effect upon the sex-ratio.

In essence, PARKES, CREW and KRALLINGER agree with the hypothesis of LENZ that the Y-bearing sperms have a greater motility
than the larger X-bearing sperms, although CREW also states that the
high primary sex-ratio may be due to a differential production of the
two types. KRALLINGER draws attention to the fact that in man, a
rise in sex-ratio is associated with a decreased fertility; this is
supported by figures. The three investigators postulate a higher
intra-uterine death rate of males. That there is such a high death
rate has been well illustrated by HAMMOND (1921) (see p. 44).

MACHENS (1915) states that there is a predominance of males in the
first litters of gilts but that after the 5th litter, females are greater in
number. He also states that the smaller the litter, the greater the
proportion of males and *vice versa*. He supports the findings of WIL-
CKENS (1886), FRÖLICH (1911) and GEORGE (1912) that more females
are born during cold than during warm weather. The data of HAYS
(1919) indicate a tendency for an excess of males in inbred litters.
PARKER & BULLARD (1913) examined 1000 litters of unborn pigs
and found the mean number per litter to be as low as 5.97. These
gave a sex-ratio of 50.64 per cent. PARKER (1914) was also able to
show that the sex-ratio is not influenced by the position of the young
in the maternal body. DASSOGNO (1915) found that the sex of the
foetus had no influence on the length of gestation period.

AXELSSON (1928) states that the difference in the sex-ratios between
the Large White and Improved Landrace is not significant, the
figures being 51.68 per cent. and 50.46 per cent. males respectively.

Kulow (1928) found a greater difference between the sex-ratio of two families, namely 97.3 per cent. and 75.0 per cent. McPhee (1932) states that in his experiments inbreeding produced a higher sex-ratio.

6. Intersexuality

There are numerous references to intersexuality in pigs, but most of the literature on this subject is confined to descriptions of individual cases and is therefore of greater physiological than genetic value.

Crew (1923a) examined two full-grown Yorkshire pigs from the same litter, one of which had cryptorchid testes and the other had more advanced male sex glands. There was a third case somewhat similar in appearance to the latter. Crew considers this state to be certainly of genetic origin and due to the mating of individuals differing in hereditary factors which direct the rate of sexual developments.

Baker (1925a, b; 1926a, b; 1928) discusses intersexuality in the pig and states that it is not very rare among both goats and pigs. In the New Hebrides, intersexual pigs are common and are found in practically every little native village. Owing to the demand for such animals for both religious purposes and for currency, they are of great importance. Many intersexual pigs that are found here differ from any of the other types of European pig intersexes in that they invariably lack any rudiment of uterus or vagina.

Baker is of the opinion that the underlying causes of the sex-intergrade in the pig and the goat are not the same as in the free-martin, in spite of the close resemblance in their anatomy. His reasons for this belief are: firstly, certain boars continually sire sex-intergrade pigs and this is not in agreement with the free-martin theory; secondly, fusion of chorions has never been found in the pig; and thirdly, sex-intergrade goats have been born singly. Baker is in partial agreement with Crew, who has suggested that such a condition may be brought about by the simultaneous development, to a certain extent, of both male and female organs. There is a definite point at which either the male or female sex organs must be encouraged to grow, and if this point be passed without differentiation, then the hormones which are produced later have not the ability to bring about a perfect development. BAKER explains that against this theory is the fact that intersexual pigs possessing a well-developed uterus have been known to exist. Again, some animals may possess both testicular and ovarian tissue. He is, however, of the opinion that the tendency towards this defect is hereditary. JAKOBIEC and MARCHLEWSKI (1932) mention the marked tendency towards intersexual forms which was observed among the offspring of an outcross of sows from the Boguchwała herd by a certain Swedish boar.

VIII. ABNORMALITIES AND DEFECTS

1. *General*

KRONACHER (1930) discusses the question of „crits", „shargars", dwarfs, backward and undersized individuals. He states that such animals are the result of genetic causes in that they possess the minimum number of factors necessary for growth and development. He also mentions what the present authors have frequently observed, namely, pigs which are perfectly normal at birth but whose development appears to be arrested, or at any rate slowed down considerably, about the age of three months. This is usually associated with a certain unthriftiness and is probably due to genetic causes.

Of all the domesticated animals, the pig is perhaps the most beset with lethal factors. Abnormal pigs pass unremarked since they occur in large litters, and therefore, so long as the abnormality does not become too frequent, the matter is not very serious from an economic point of view. A monotoccus animal producing an abnormality is immediately marked. Not so with the polytoccus pig. It is difficult to distinguish in the pig whether these not uncommon abnormalities are genetical in their origin or due to some damage in utero or during parturition.

2. *Defects of the Skull*

Amongst the commonest of such reported abnormalities are those which relate to the skull and the skin. NORDBY (1929*a, b*; 1930) describes some interesting defects of the skull. One defect he describes as „of the meningocoele and proencephalus types representing incomplete development of the neural tube involving the bony investment surrounding the brain and especially the frontal and parietal bones." This abnormality is of the nature of a brain hernia on the front part of the skull. Over 200 cases were studied, chiefly in the Berkshire and Duroc-Jersey breeds, many having come to light during inbreeding. Sufficient evidence has been secured to show that this defect is definitely hereditary, but the details concerning the mode of inheritance have yet to be published. HUGHES and HART (1934) observed four pigs with a similar skull defect in the Poland-China herd at the California Experiment Station. In each case there was an opening in the median line of the skull, associated with the parietal bones, varying in length medially from 6 mm. to 16 mm. and in width laterally from 4 mm. to 6 mm. The abnormality appeared in very closely inbred stock. These writers conclude that the defect is inherited and depends upon a recessive factor.

3. *Defects of the Ears*

In the defect described above, NORDBY noticed that the ear is also affected; the defects in the external ear may vary from a slight modification of the anterior border to a very pronounced reduction in size. The present writers have observed similar defects in the pigs in Great Britain, but have not had an opportunity to study the genetics of the condition; it appears to them, however, to be associated with sterility. Idaho Agricultural Experiment Station (1931) reports cases of dwarfed and absent ears which are quite common in one strain of Duroc-Jerseys, and can be traced to an earless dam of a famous show boar. Affected specimens also revealed a number of skull defects.

4. *Cleft Palate*

The occurrence of cleft palate is reported by M'PHEE (1932). SCHOTTERER (1933) also reports cases of cleft palate which were present in a litter from a sow who had previously produced 3 healthy litters. Among 12 piglings, 3 exhibited various grades of cleft palate and jaw, while another 3 had facial defects. The boar was normal and had sired healthy pigs. The author excludes the possibility of mechanical disturbance and suggests a genetic basis. KOCH & NEU-MÜLLER (1932) observed cleft jaw in 6 litters all sired by the same malformed Berkshire boar. This boar was only bred once to 6 sows at 6 months old. He sired 57 piglings, the majority of which were also hare-lipped. All the malformed animals were either born dead or died shortly after birth. In some cases, hare-lip was connected with ab-normalities such as lack of anus and tail. These 6 sows when mated to normal boars gave normal offspring. This defect was not sex-limited. Results indicate it to be genetic in nature and transmitted by the sire.

5. *Defects of the Eyes*

KOSSWIG & OSSENT (1932) report a case of congenital blindness and add that they are unable to explain the mode of inheritance of this defect.

HALE (1933) describes a case of a Duroc-Jersey gilt who farrowed eleven pigs, all born without eye balls. Ten were alive at birth, but all died within five minutes. It must be noted that the gilt was re-ceiving a vitamin-deficient ration. From the results of further breeding tests with the sire of these pigs, the author is inclined to suggest a nutritional rather than a genetic cause.

One of the present authors (A.D.B.S.) has been collecting instances of eye abnormalities which have occurred in various herds. The ab-normalities range from complete absence of eyes and total blindness to abnormally protruding eyes. Some of the animals lacking eyes have survived and reached slaughter or breeding age, but so far no animals with protruding eyes have been known to live for any length of time. Two pedigree and slightly related Middle White sows farrowed litters by the same boar. Both litters were abnormal, some of the piglings being totally blind whilst others had large protruding eyes. The following litters, by another boar, were normal. Another similar instance is that of a litter by a Small White boar out of a sow who was his sister. The entire litter was abnormal while the preceding litter of the same sow had been normal.

There are other descriptions of similar cases in which several or all of one litter of a sow were partially or completely blind. Two other cases worthy of particular note are: 1) a Large White sow who farrowed

13 pigs of which three were blind. The dam of this sow also had two blind offspring among her litter; 2) another sow had a litter of 17 in which there were two dead and the remainder completely or nearly blind. The mother of this sow had a still-born litter by the same boar.

From these data it would not appear entirely unreasonable to suggest that certain forms of blindness are hereditary, since a small degree of inbreeding is often associated with this defect. Other cases reported are probably due to damage in utero.

6. *Tassels*

It is not uncommon for the pig to possess tassels like those of the goat. They are found occasionally in sheep and rarely in man. Though they do not occur in the recognised breeds, they are not infrequently met with amongst the pigs of Europe, from Spain to Russia. One of the earliest mentions of this abnormality appears to be that of EUDES -DESLONGCHAMPS (1842), (as quoted by DARWIN (1868)). He stated that the appendages which often characterized the Normandy pigs were always attached to the corners of the jaw, three inches in length, cylindrical and covered with bristles. The centre was cartilaginous, with two small longitudinal muscles. LUSH (1926) reports their occurrence among the unimproved swine of the southern United States. KRONACHER (1924) has investigated their inheritance. Like those of other animals, these tassels or „bells" consist of a tongue of cartilage (absent in the sheep) supplied with vessels and nerves, and of connective tissue. In the pig, the neck tassels are also supplied with muscle. His evidence points to tassels being a simple dominant factor. KRONACHER also mentions an aberrant bell which he found at the root of the posterior surface of both ears in an Improved Landschwein boar. The boar was crossed with several sows of the same breed, but among 100 F_1, only one had two bells at the root of the same ear. It is worth noting that tassels in the goat are inherited as a simple dominant (ASDELL and BUCHANAN SMITH 1927). Aberrant tassels in the caprine species are almost certainly genetic in origin but are almost as aberrant as the species.

7. *Hairlessness*

Opinions are diverse as to the cause of this anomaly. Earlier workers suggest malnutrition in the form of iodine deficiency and give examples of hairless pigs which have been cured by some form of iodine treatment. On the other hand, the results of more recent work are definitely in favour of a genetic explanation in certain cases where iodine feeding has failed, and there is evidence to show that the hairless condition has been brought about by breeding methods.

SMITH and WELCH (1917) draw attention to the common occurrence of this defect in many parts of the north-western States of America. In some districts it is serious and many animals die. They suggest that the cause is lack of iodine in food and water, and find that the administration of potassium iodide to pregnant females gives good results. Post mortem examination of hairless individuals revealed hypertrophy of the thyroid, an under-developed heart, and thick pulpy skin. Chemical examination showed an extremely low iodine content of the thyroid. Since one of the most marked characteristics of the abnormality is the absence of hair, it may be assumed that there is probably a direct relationship between the physiologically active secretion of the foetal thyroid, and growth of the epidermal appendages. HART and STEENBOCK (1918a, b) support the view of SMITH and WELCH (1917) that hairlessness is caused by low iodine assimilation, which results in a goitrous condition in both mother and young and interferes more severely with the development of the foetus than with the mother. Sows which produce normal offspring can have this condition induced by feeding rations with a high protein content. It is suggested that rather than feed iodine, more attention should be paid to the combination of natural materials. SHEPPERD et alii (1924) report a case of 4 pure-bred Duroc-Jersey sows which were fed 6.5 grs. of iodine daily for 18 days before parturition but in spite of this, 31 hairless offspring were produced. However, a grade sow having received iodine for 2 weeks longer, farrowed 7 haired pigs.

ROBERTS and CARROLL (1931) state that hairlessness is common among Mexican swine where it cannot be prevented by the administration of iodine. As the result of crossing grade Chester Whites with two males and two females imported from Mexico, it was thought that the normal condition must be incompletely dominant. The normal condition was designated H, and the abnormal h, and by further experiments it was assumed that one gene was involved. This is in accordance with ROBERT's work of 1925 and 1927. By 1928, as a result of matings of heteroyzgous parents, ROBERTS had obtained 909 haired and 293 hairless individuals. Histological examinations of the thyroids and adrenals of these animals revealed no abnormalities. When hairless and haired swine were mated, the F_1's possessed coarser hair and a thinner coat than the normal animals. Further studies were carried out by SEVERSON (1932).

DAVID (1932) examined 3 specimens of swine skin taken from a normal pig, a heterozygous hairless pig, and a homozygous hairless pig. The animals homozygous for hairlessness showed a few hair follicles, while the heterozygotes showed an intermediate condition. In the skin of the homozygous hairless, there was a definite tendency towards irregularity in the direction of the follicles which were otherwise structurally normal. The coat of this specimen was much shorter, as many of the hairs were broken off. The number of sweat glands was reduced and this corresponded to the reduction in the number of hair follicles; the sebaceous glands were rudimentary.

8. *Defects of the Skin*

NORDBY (1929b) reports a case of skin abnormality which is probably genetic. The defect consisted of five more or less circular areas in which the epidermal layers were incomplete, in some cases the aponeurosis being exposed. In another paper (1933a) he describes the occurrence of melanotic skin tumours in pigs. In this instance, a boar with a „wart" on his loin was purchased for breeding purposes. In all of the first five families of pigs sired by this boar, this trouble appeared, 10 or more affected animals being produced in the first generation. Later, he was mated to his affected daughters with the result that in 3 litters, 8 pigs with „warts" appeared. Considering the frequency of the occurrence of this defect, the author is of the opinion that it must be inherited but not as a simple dominant. Owing to the economic loss involved, (hams with this blemish were not passed as first class by United States inspection), strict selection and the elimination of affected animals is advised.

9. *Seedy Cut or Black Belly*

The cause of seedy cut is said by DEAKIN (1932a) to be due to the presence of black pigment in the mammary glands when it is either present in the skin, or else absent, as in the Duroc-Jersey. The time at which the pigment appears in the nipple epidermis and the ability of the epithelial cells to synthesise pigment, control the presence or absence of pigment in the glands of the black breeds. The ability of the cells to synthesise pigment and its occurrence are correlated with intensity of pigmentation and the rapidity of cell division. DEAKIN expresses doubt as to the possibility of evolving a strain of pigs which would be entirely free from pigmented glands. Selection can be practised in red pigs, in which the glands are visible through the skin of the live animal, but in other colours this is not the case. ARMSTRONG (1932b) advises the use of white boars as a preventive measure.

MACKENZIE and MARSHALL (1915) undertook experiments to ascertain the possibility of destroying mammary pigment in cells during the period of glandular activity. Three Large Blacks and one Berkshire were used and pigment was detected in practically all the glands operated upon. After an interval of 17 months, during which period each of the sows had two litters, the animals were killed. The glands were re-examined and in no case was pigment detected.

COLE, PARK & DEAKIN (1933) describe the two distinct types of „seed". One is due to the presence of pigment in the mammary gland (as described above), and the second to vascular hypertrophy and may be red, pink or white. All gilts seem to exhibit some phase of this type of „seed". It appears in the gland when the gilt reaches maturity; it is red immediately following oestrus, and changes through pink to white during dioestrus. The seed increases with the number of cycles. Barrows (castrated males) are not affected by this vascular type of seed. Black seed (type 1) occurs in half the gilts of black breeds, in 23 per cent. of those of black and white breeds, and in 20 per cent. of those of red breeds, but not at all in white breeds. It occurs in 68 per cent. of the barrows of black and black and white breeds, but not at all in barrows of red or white breeds.

10. *Scrotal Hernia*

According to WARWICK (1926b), the earliest reference to hernia in swine appears to be that of YOUATT (1847). WARWICK mentions the following writers as being in favour of the hereditary nature of this defect: CAMPBELL (1914), FLEMING (1902), HOBDAY (1914), JOEST (1921), LEENY (1920), MÖLLER & DOLLER (1903), and WHITE (1914). WARWICK himself (1928a) has shown that scrotal hernia in inherited in a recessive manner. By definitely selecting for this defect, he was able to increase the percentage of ruptured animals per generation at the following rate: 7.49% — 14.28% — 42.0% — 39.4% — 47.5% — 45.9%. He states that 1.73 per cent. of all pigs have hernias. Practically all the ruptured females had umbilical hernia, but this type was rarer in the male.

Later (1931) he writes: „Hernias, or ruptures, of swine are so common that scarcely a swine raiser has escaped experience with them. The two most common kinds are scrotal, or inguinal, hernia and umbilical or navel hernia. Scrotal hernia consists of an enlargment of the scrotum by loops of bowel. Although the anatomical differences of the sexes necessarily limit the occurrence of scrotal hernia to the male, females sometimes are seen which have inguinal hernia, which would be comparable. So far as our observations go, we have no reason to believe that there is any hereditary relationship between the two. The loops of bowel pass through the opening of the abdominal wall (inguinal canal) with the spermatic cord which connects with the testicle. Umbilical, or navel, hernia is formed by loops of intestine passing through the abdominal wall at the umbilicus, or navel, but without a break in the skin.''

Breeding tests to determine whether scrotal hernia is heritable in swine were initiated by WARWICK at the Wisconsin Experiment Station in 1922. They were carried on in co-operation with the U.S. Department of Agriculture until 1926, when they were transferred to the Ohio Experiment Station and there continued until 1929. The fuller data were published by WARWICK in 1926. Scrotal hernia occurs more frequently on the left side than the right. It is never present at birth and rarely occurs after one month of age: it may appear at any time from one day to one month. The following table from WARWICK (1931) gives the figures he obtained: (see table page 68).

Further figures clearly show that there is a tendency towards the occurrence of scrotal hernia in the male pig.

WARWICK (1928a, 1931) advances the hypothesis that the condition is dependent upon two pairs of recessive factors, h and h', in a homo-

Summary of the Occurrence of Hernia in Pigs raised to one month of age in the Experimental Herd.
(From WARWICK 1931, p. 26)

	Generation of Selection for scrotal hernia							Total
	1	2	3	4	5	6	7	
Number of pigs	132	163	122	125	106	24	7	679
Number herniated	11	41	34	32	29	11	2	160
Per cent. herniated	8.33	25.15	27.86	25.60	27.34	41.66	28.57	23.56
Number of male pigs . . .	66	92	69	70	57	11	5	370
Number males herniated .	11	39	31	31	27	10	2	151
Per cent. males herniated .	14.28	42.39	44.92	44.28	47.37	90.90	40.00	40.81
Per cent. males scrotally herniated	14.28	42.39	43.47	44.28	47.37	81.81	40.00	40.27
Number males umbilically herniated	0	0	1	1	0	1	0	3
Number females umbilically herniated	0	2	2	0	2	1	0	7

zygous condition. The affected boars would thus be $hh\ h^1h^1$, while sows of the same genotype would be normal. The evidence he adduces is most convincing, despite the fact that no certainly homozygous sows were used in testing it.

During an attempt by McPHEE (1932) to inbreed various strains of different breeds, hernia arose when it was not known to exist in the foundation stock.

11. *Cryptorchidism*

McKENZIE (1931) found evidence that cryptorchidism was inherited and mention is made by NORDBY (1933b) of the prevalence of this hereditary character of cryptorchidism in swine. The extent to which this defect is prevalent may be seen in the table below in which he has summarised the observations of several workers. NORDBY strongly advises that cryptorchid boars, or those which have sired cryptorchid pigs, should not be used for breeding.

Summary of numbers and percentages of cryptorchids reported by BUSMAN, DE WOLF, JELEN and SHELTON. (NORDBY 1933, p. 902)

Investigator	Hogs Observed	Males	Crypt-orchids	Cryptorchids on basis of total males %
BUSMAN (Chicago) . .	103,000	49,018[1])	313	0.64
DE WOLF		4,671	35	0.79
JELEN (Omaha) . . .	534,486	254,312	2,138	0.84
SHELTON (Denver) . .	142,000	67,578	493	0.73
Totals	779,486	375,579	2,979	

Among a population of 107 pigs sired by one boar out of 12 different sows, McKENZIE (1931) found ten cryptorchids, which came from five sows. Of these sows, four were related. The sire of these ten pigs was again mated to his daughter who was a litter mate of one of the affected animals. Of the resulting progeny, about 50 per cent. of the males were cryptorchid, and from these facts McKENZIE concludes

---

[1]) The figures for males have been derived, where necessary, by taking 47.59 per cent. of the total number of hogs. In the sex classification based upon figures in the U. S. Department of Agriculture Year Book, 1930, Table 384, barrows comprised 47.59 per cent. of the total number of hogs from 1923 to 1929 inclusive.

that this defect must be hereditary. McPHEE and BUCKLEY (1934) conclude that cryptorchidism is a sex-limited recessive character but are unable to determine more exact details of its inheritance. Their results show that inbreeding tends to increase the proportion of cryptorchids and they emphasise the inadvisability of using a cryptorchid boar, should he prove to be fertile. These authors support the conclusions of earlier writers, that all boars and sows which are known to have produced cryptorchid offspring should be eliminated from the breeding stock.

12. *Atresia ani*

KINZELBACH (1931) made observations on this defect in the Swabian-Halle breed. In males the defect is characterized by a complete absence of the anus and anal cavity while the rectum is shortened. Affected animals may be kept alive by the construction of an artificial anus by operative means; the success of the operation depends upon the extent of the rectal malformation. Animals upon which this operation has been successfully performed develop normally. In females there is also complete absence of the anus and the rectum is defective. Defaecation, however, proceeds through the recto-vaginal passage without any difficulty and the affected animals develop more or less normally. Generally, they are able to breed.

During the period 1927—29, out of 621 animals distributed in 36 herds, KINZELBACH observed 103 affected cases, 86 males and 17 females. He noted that the phenomenon seems to be exhibited by the progeny of definite sires and dams and undertook breeding experiments in order to obtain further data. The results obtained were as follows:

(i) *Normal* × *Affected*

Sire	Dam	Progeny					
		Litters	Total	Normal		Affected	
				♂	♀	♂	♀
Affected	Normal	2	13	4	6	1	2
F₁ normal	F₁ normal	3	16	5	10	1	—

(ii) *Affected × Affected*

Sire	Dam	Progeny					
		Litters	Total	Normal		Affected	
				♂	♀	♂	♀
Affected	Affected	2	11	2	5	2	2
F_1 normal	F_1 normal	1	6	3	2	—	1
*F_1 affected	F_1 normal	2	23	11	7	3	2
*F_1 affected	F_1 affected	1	8	3	2	2	1

The material consisted of three affected boars (which had been operated upon), three affected sows and one normal sow. Only one of the affected and the normal sow became pregnant, both the other sows having to be discarded.

The author concludes from his study that this defect is inherited and that, among the progeny of affected parents, there occur individuals which, though homozygous for the defect, do not manifest it. He points out that this type of inheritance has been observed in other forms, and suggests that recent observations on the pig by other writers and himself might be interpreted in this manner.

WALTHER, PRÜFER and CARSTENS (1932) submit evidence that the factors for *atresia ani* and for thick forelegs are linked *(vide infra)*.

13. „Kinky" Tail

NORDBY (1934b) describes a defect in the tail of swine which he calls „kinky" tail and which is genetic in its origin. The defect is characterised by rigid angles in the tail and is caused by unilateral fusion of adjacent caudal vertebrae. From the data accumulated, NORDBY deduces that this defect is due to a single recessive factor acting in the presence of inhibitory influences, so that the normal ratio does not appear.

14. *Paralysis of Hind Limbs*

MOHR (1930), cited by HUTT (1934), reports a condition character-

---

*) The same boar.

ised by complete paralysis of the hind limbs in homozygous piglings. The animals failed to live unless subjected to special treatment; the character is, therefore, a lethal one. When heterozygotes were mated together, normal and paralysed piglings were produced in the ratio of 71 : 25, from which the author deduces that the condition is dependent on a single gene.

15. *Thick Forelegs*

WALTHER, PRÜFER and CARSTENS (1932) describe this phenomenon which consists of a gelatinous infiltration of connective tissue and modification of the normal muscle tissue through this connective tissue. The bones are also considerably thickened. Great variation occurs in the degree to which the abnormality is exhibited and it appears to be lethal in all cases except where the defect is only a slight deviation from the normal condition. Matings of a boar which exhibited this abnormality (and also carried factors for *atresia ani* and congenital blindness) with three of his daughters which carried the factor for thick forelegs gave the following results:

No. of Sows	No. of litters from mating with the sow's sire	No. of piglings	With thick forelegs	Percentage of total number	With *atresia ani*	Percentage of total number	Both thick legs and with *atresia ani*
1	3	26	6	23	3	12	2
2	2	14	5	36	2	14	2
3	1	11	2	18	2	18	1
	6	51	13	25.5	7	13.7	5

The authors point out that in the case of thick forelegs the normal ratio (25.5%) of abnormal offspring was obtained, but in the case of *atresia ani* only 13.7% abnormal offspring were obtained in place of the expected 25%. This agrees with the findings of KINZELBACH (1931), (*vide supra*). With regard to linkage between thick forelegs and *atresia ani*, they are of the opinion that since 5 out of 7 piglings with *atresia ani* (71%) also exhibited thick forelegs, this constitutes undeniable evidence of linkage since the number of animals exhibiting both conditions was nearly three times as great as would be expected without linkage.

16. *Stringhalt*

WARWICK (1931) has observed at both the Wisconsin and the Ohio Experiment Stations pigs which in walking, jerk their hind legs. The feet may be jerked as high as the back. A partial recovery usually takes place after several months. The pigs affected at Ohio were all sired by one normal boar and two affected daughters were bred back to him. One of these gilts raised four pigs all of which showed abnormal locomotory powers. The defect occurred when they were only a few weeks old, but later they made a practically complete recovery; this was in marked contrast to the affected pigs observed previously, none of which showed the trouble at an earlier age than 14 weeks. The inheritance of these defects is considered to be due to a combination of recessive factors.

17. *Bent Legs*

The occurrence of „bent legs" in the Swedish Large White is described by HALLQVIST (1933). The legs are said to be stiff and bent at right angles to the longitudinal axis of the body. Either the fore-legs alone or the hind legs as well, are affected. The animals are usually still-born or die very soon after birth. In 32 litters, 220 piglings were normal and 46 abnormal which could all be traced to one boar. When 15 sows and 1 boar were tested for heterozygosity by mating with known heterozygotes, 10 sows proved to be heterozygous and the remainder homozygous or normal.

18. *Knock-Knees*

EVANS (1930) states that, as the bacon breeds are becoming higher off the ground, the animals have to spread their forelegs in order to eat and this no doubt causes a tendency to knock-knees. This defect, however is more marked in some strains than in others and appears to be most certainly of genetic origin.

19. *Syndactyly*

ARISTOTLE, who related that „in Illyria, Paeonia and other places there are swine with a solid hoof", appears to have been the first to report syndactylous pigs. The classical case of these solid-hoofed pigs has been described by STRUTHERS (1863) on the estate of NEIL MENZIES at Rannoch, Scotland. Most of the animals were black. The original pair was brought to Rannoch about forty years previously. They increased to several hundred and one wonders what the Scottish Highlander with his instinctive religious (and hereditary?) dislike of the porcine species thought of them. At the time STRUTHERS wrote, they were practically extinct. DABROWA-SZREMOWICZ (1905a) reported similar pigs with „ a median line of demarcation of the hoof" which originated from a sow whose unusual description included „a long tail ending in a tuft". DARWIN (1868) gives a picture of an Irish pig which answered well to this description, and also possessed tassels.

One of the earliest reports is that of the veteran investigator of animal peculiarities of Europe and America, AULD (1889), who states that „soliped" pigs were reported from Texas in the year 1878. An anatomical description is followed by the statement that the breed was so firmly established that no tendency to revert to the original form was observable. In the cross of solid with normal-hoofed animals, the majority of the litters were solid-hoofed. AULD indicated that „mule-footed" hogs were of frequent occurrence and cited cases in Iowa and Louisiana. Such pigs are not uncommon in America at the present day. The SIMPSONS (1908) considered the mule-foot to be a simple dominant over normal hoof. DETLEFSON and CARMICHAEL (1921) mated a black mule-foot boar to Duroc-Jersey sows. 280 individuals were raised, all of which were syndactylous and self black. Among the F_1 females by a Duroc-Jersey boar, the ratio of mule-foot to cloven-foot was 17 : 23. There appeared to be no linkage of this defect with colour.

MALSBURG (1924), with Polish pigs, and KALUGIN (1925), with Russian pigs, in matings of solid-hoofed × solid-hoofed obtained respectively ratios of 19 solid-hoofed : 7 normal and 55 solid-hoofed : 13 normal. The condition is thus a simple dominant. MALSBURG observed three individuals that were syndactylous on the forefeet only; these are not included in the above ratio. KRONACHER (1924) reports a case of a boar out of an Edelschwein sow × Improved Landschwein boar,whose left hind foot was solid. All his 22 offspring out of normal sows were normal.

20. *Polydactyly*

This condition is met with in the pig as in other animals and has been known for many years. BATESON (1894) states that „of the great numbers of polydactyle feet recorded or preserved in museums all, I believe, are fore feet. No case of a polydactyle hind foot is known to me in the pig. All the cases are examples of proliferation upon the internal side of the digital series".

MEINERS (1922) reviews the whole subject and adds a description of 7 cases noted by himself. In 92 cases, polydactyly was unilateral and in 7 cases, bilateral (forefeet only). The fact that in 93.34 per cent. there was anterior polydactyly and in 95.34 per cent. median polydactyly, is thought to be of special interest. He states that most of the cases hitherto described represented teratological phenomena and were mostly caused by disturbances of the amnion during development.

In matings of polydactylous pigs, KALUGIN (1925) used one boar and six sows; only two of these sows showed complete polydactyly, i.e. not fewer than five toes on each foot. 53 offspring were produced, of which 45 were polydactylous, 32 being completely so and 13 showing 4 or more digits on one or more feet. The normal-toed animals numbered 8. KALUGIN considers the condition to be a Mendelian dominant. In matings of syndactylous by polydactylous animals 29 young were obtained. Four types were present in the following numbers: normal 4, syndactylous 3, polydactylous 10, combined syn- and polydactylous 12. These figures represent 22 polydactylous: 7 non-polydactylous, a 3 : 1 ratio, and 15 syndactylous to 14 non-syndactylous, a 1 : 1 ratio. The results are interesting but scarcity of numbers barely justifies otherwise reasonable speculation.

SCHOTTERER (1933) obtained among a litter of 12 piglings, 6 showing polydactyly of the anterior limbs, the number of toes being 4, 5, or 6. This defect was associated in some instances with cleft palates. HUGHES (1934) observed 13 cases of polydactyly in Duroc-

Jersey pigs in a population of 125 at the California Experiment Station. The extra toe occurred equally among males and females and on one or both forefeet; the condition was not observed on the hind feet. The expression of the abnormality was so irregular that no attempt was made to analyse the mode of inheritance.

21. *Two-legged*

DARWIN (1868) quotes Colonel HALLAM who described a race of two-legged pigs, „the hinder extremities being entirely wanting". This deficiency was transmitted through three generations. The pigs were observed „at a town on the coast in the Tanjore country, in the year 1795". One wonders what BUFFON would have said to these variations in the limbs of the pig, an animal which he describes (1780) as „an ambigious species" and later adds, „All his habits are gross: all his appetites are impure; all his sensations are confined to a furious lust and a brutal gluttony" [1]).

That „the animal participates of several species" he ascribes largely to its feet, „Its extremities which are cloven-hoofed have no resemblance to those which are whole-hoofed. It even resembles not the cloven-hoofed animals; because, though it appears to have only two toes, it has no resemblance to the digitated quadrupeds; because it walks only on two toes, and the other two are neither so situated nor extended so far, as to serve the purposes of walking. It has, therefore, equivocal or ambiguous characters."

IX. ANATOMY AND CONFORMATION

1. *Face and Skull*

The action of man in the moulding of the face of the animals which he has domesticated, has, with the exception of the dog, nowhere had so great an effect as in the pig. The face of the pig may be of extreme length, like that of the deer hound (Saluki), or it may be almost as dished as that of the bulldog or pug.

---

[1]) Contrast this with Buffon's description of the goat: „The he-goat is a beautiful, vigorous and ardent animal."

The SIMPSONS (1909) found that the long-snouted Tamworth crossed to the Large White (the latter having somewhat shorter faces than is the present fashion) gave progeny with an intermediate face shape. WENTWORTH and LUSH (1923) crossed the extremely long-faced Wild pig with the brachycephalic Berkshire. Save for slightly wider foreheads, the two F_1 which matured were indistinguishable as regards head shape from their Wild parent, nor could they be distinguished in this respect from the F_1 of the Wild × Tamworth. The same writers report that the mating of Tamworth × Berkshire produced seven pigs with the typical Tamworth face, which is contrary to the results of the SIMPSONS. WENTWORTH and LUSH therefore conclude that in spite of the Berkshire and Yorkshire having a similar phenotype in this respect, their factor complexes are not identical. They also worked with the Duroc-Jersey which is intermediate between the Yorkshire and Tamworth. The face shape of the F_1 from the Duroc × Berkshire cross approximated to that of the Berkshire but was not sufficiently dished to be the ideal type. In the F_2, there was considerable variation, the results obtained being as follows:

Facial character	Similar to Berkshire	Intermediate	Similar to Duroc-Jersey
Forehead shape . . .	37	2	3
Dish of face	17	9	16
Length of face. . . .	21	7	14

NATHUSIUS (1864) found that in the early-maturing pigs, the skull is relatively broad and deep, but in the late-maturing animals the skull is long and narrow, as that of the Wild boar.

KRONACHER (1930) found that the males of Wild × Landschwein cross had a skull intermediate though approximating to that of the Landschwein, while the sows approached the wild type very closely, but the evidence here is not good. TEODOREANU (1929) also made some observations concerning the profile and length of snout. CONSTANTINESCU (1933) is of the opinion that in the F_1 of Mangaliţa × Middle White, the Middle White profile is recessive and the head length intermediate.

HABU (1930) investigated the skull measurements of Middle White, Berkshire, Large White, Poland-China and Duroc-Jersey breeds and their F_1. The following measurements were used for purposes of comparison — the length: breadth index of the cranium and the curvature of the profile (the magnitude of the angle formed by the middle of the cranial line, the middle of the fusion of the frontal and nasal bones and the most anterior point of the nasal bone). The results indicated that the concavity of the line of profile is independent of sex and increases with age. The breeds are classified according to the average value of the angle, i.e. 138°—142° for Middle White and Berkshire, and 160°—164° for those with a straighter profile, i.e. Large White, Poland-China and Duroc-Jersey. In crosses between these breeds, and their reciprocals, the concavity behaves as an intermediate, with the exception of the Berkshire × Poland-China in which case the influence of the sire appears to be greater. Sex appeared to have no influence upon the length : breadth index, but again with one exception, that of the Yorkshire × Duroc-Jersey, in which the male appeared to have a greater influence.

2. Ears

According to CARR-SAUNDERS (1922), the erect ears of the Berkshire are dominant to the lop ears of the Large Black. WENTWORTH and LUSH (1923) describe the result of crosses between the Berkshire and the Duroc-Jersey. The ears of the latter breed, while not lop, are of medium size, less pointed, and the outer third breaks over sharply and droops downward. The F_1 conformed more closely to the Berkshire than to the Duroc-Jersey, while the majority of an F_2 of 42 were like the Berkshire, a few were intermediate and one had a typical Duroc-Jersey ear. WENTWORTH and LUSH think that there may be as many as three factors involved, as well as possibly a number of modifying factors for size, quality, and the amount of breaking over of the ear.

KRONACHER (1924), working with parental breeds that were less homozygous than the above, comes to the conclusion that ear conformation is based on a series of factors, and that the position of the ear is independent of ear shape. One animal had one lop ear while the other was „semi-erect". On the whole, the tendency was for the F_1

to be intermediate but with variation to dominance occurring in both directions according to the parental breeds employed. For instance, the tendency was for the lop ear of the Cornwall (Large Black) to be dominant over the more erect ear of the Improved Landschwein but only incompletely dominant over that of the Half-red Bavarian Landschwein.

CONSTANTINESCU (1933) states that the form and direction of the ears of the F_1 of Mangaliţa × Middle White crosses are indicative of incomplete dominance of the Middle White type.

Of interest is the observation of DARWIN (1868) who stated that the pigs then found in the Orkney Islands were small, with erect and sharp ears and „with an appearance altogether different from the hogs brought from the south."

3. *Body, Length and Ribs*

PLAIM (1930), from measurements taken from a limited number of pigs of various breeds and crosses, states that on the whole, the pelvis is narrowest in the primitive European domestic pig, broader in the curly-haired cross of European and Indian pigs, and broadest in the highly-bred English breeds and their crosses.

CONSTANTINESCU (1933) found the body length of the F_1 of the Mangaliţa × Middle White cross to be intermediate between that of the parental types. FERRIN (1933), as a result of work carried out at the Minnesota Station, is of the opinion that depth and width of body are pronounced in the early-maturing type of pig.

SHEPPERD et alii (1924) state that a bacon pig has 15 pairs of ribs whilst a lard pig has 14 pairs. In a group of 90 pigs, they found that 5 had 14 pairs; 60 had 15 pairs; 20 had 16 pairs; and one had 16 on one side and 17 on the other. A more detailed investigation was undertaken by SHAW (1929). From some 4,000 records gathered from many different breeds and crosses, he obtained a definite degree of correlation between rib number and the „placing" of the carcase. The number of ribs varied from 13 to 17 pairs. Though variation occurred in all breeds, there was a distinct breed difference in the number. The larger breeds, especially those of the coarse type, appeared to have a greater number of ribs. The following table is of interest:

Table showing numbers of ribs occurring in different breeds, grades and crosses. (From SHAW 1929, p. 25)

Breed	13	13+*	14	14+	15	15+	16	16+	17	
Yorkshire			94	2	393	14	169	1	5	678
Berkshire	1		13		2					16
Tamworth			12		11					23
Duroc-Jersey	1		175	24	246	13	25			460§
Poland-China	1	2	318	63	397	23	33			837
Chester White	4	6	314	70	231	10	6			641
Hampshire	1		121	11	120	9	13			275
Yorkshire† × Tamworth .			1	1	5		2			10§
Iowa Grades	1		54		40		5			100
Large White			34		7		1			42
Middle White			40		21		2			63
Large Black			2		9					11
Gloucester Old Spots . . .					12					12
Welsh			16		36		2			54
Lincolnshire Curly-Coated .			34		7					41
Large White × Large Black			5		10					15
Large White × Berkshire .			21		8					29
Large White × Middle White			17		19		1			37
Oxford × Large White . .			17		13					30
Tamworth × Berkshire . .			35		13					48
Large White × Welsh . . .			7		6					13
Welsh × Large White . . .			22		37		10			69
Tamworth × Gloucester Old Spots			8		4					12
Tamworth × Yorkshire . .			6		10		3			19
Berkshire × Tamworth . .			5		6					11
Berkshire × Yorkshire . . .			4		7					11
Yorkshire × Landrace . .	11		185		139		37		2	374

Rib number appeared to be closely correlated with length of side of bacon and thus with economic production, but there seemed to be no connection between sex and rib number. Litter mates showed less variation than unrelated animals.

---

*) In 2nd, 4th, 6th and 8th columns are listed pigs showing uneven pairs or where one or more ribs were „floaters" or in any way defective.

†) In the case of crossbred pigs (×) the breed of the boar is stated first.

§) So in original.

Table showing variations among litter mates (From SHAW 1929, p. 26)

No. of litter	No. of pigs in litter	Rib No.						
		14	14+	15	15+	16	16+	17
I	13	6	1	6				
II	12	5		5	1	1		
III	12	1		11				
IV	12	1		11				
V	ł2			6		5		1
VI	10	3		6		1		
VII	8					7		1
VIII	14	3		8		3		
IX	10			3		7		
X	12			9	1	2		
	115	19	1	65	2	26		2
Percentage		16.52	0.87	56.52	1.74	22.61	0.0	1.74

AXELSSON (1933) obtained correlations of $r = + 0.29$ and $+ 0.35$ for body length and number of ribs. Although no genetic studies have been made concerning rib count, this quality is almost certainly principally governed by heredity. The development of a strain of pigs of uniform rib number should have possibilities, more especially if this character can be relied upon as an indication of carcase value.

Variation in body length has been investigated at several pig testing stations, especially those of Sweden where attention has been directed towards this aspect of conformation owing to the relationship between body length and thickness of back fat. (*See* „Productive Qualities"). From their results, HANSSON and BENGTSSON (1926, ff.) demonstrate the importance of selecting for breeding stock sows which have long sides. In Denmark, JESPERSON and MADSEN (1929) confirm this statement. They also obtained a definite degree of correlation between back length and thickness, which cannot be greatly affected by either nutrition or environment.

The relationship between conformation and market value has also been emphasised by FERRIN (1933) and HAMMOND (1932, 1933). The latter draws attention to the breed difference in the rate of development and growth. Certain breeds, such as the Middle White, reach their maximum body length at a comparatively early age whereas other breeds, e.g. Large White, require a longer period of time.

It would appear to the writers that body length must be very closely connected with rib count and therefore adds even greater value to rib number investigations.

X. PRODUCTIVE QUALITIES

1. *General*

The destiny of the pig is meat. The type and quality of the meat is determined by the appetite of the human consumer whose demands are directed along three channels: fresh pig meat or pork, cured pig meat including bacon and hams, and lard; in this last case the main function of the pig is to supply fat. The fact that certain parts of the pig can be utilised in an other form, e.g. as sausages, does not imply that such a commodity is a principal product. Generally speaking, sausages etc., are an important by-product of pigs produced either for bacon or for pork. Within wide limits, the type of the pig has no great effect on the quality of the sausage.

The type of the pork pig varies with the market, from the sucking pig of some 40 lbs. (18 Kg.) live weight up to or over 200 lbs. (90 Kg.). As a rule, the range in most countries is from about 80 to 140 lbs. live weight (36—64 Kg.), though in the United States 225 lbs. (102 Kg.) is considered to be the optimum weight for fresh pig meat (the term „pork" is not used in the United States). The pork pig must be a quickly-maturing animal and a reasonable amount of fat is no objection.

The minimum live weight of the bacon pig is 190 lbs. (86 Kg.), with a maximum of 300 lbs. (136 Kg.) in certain countries and districts; as a rule the optimum ranges from 195 to 240 lbs. (88—109 Kg.). The English market for bacon, cured in what is known as the Wiltshire manner, requires pigs from 195 lbs. to 220 lbs. (88—100 Kg.) live weight. Such a pig usually kills to about 75% of its live weight. The dead weight of such a pig should be not less than 140 lbs. (seven score) (64 Kg.) and not more than 170 lbs. (77 Kg.). Since the English are the biggest buyers of pigs in the world market, these are the weights to which the producers of all exporting countries must conform.

Although the United States export a great quantity of pig products, consumption is over 90% of production. The type of bacon hog is slightly heavier than that required for the English Wiltshire market and ranges from 210 to 250 lbs. (95 to 113 Kg.) live weight. In North America, the same type of pig suits both the bacon and fresh meat trades.

The third type is usually known as the „lard" type of hog. Such pigs weigh from 300 lbs. (136 Kg.) upwards live weight. Their major function is the production of fat, i.e. lard, of which, till recently, the United States have exported large quantities to the continent of Europe. The belly meat of these animals is utilised to provide the dry salt cure (PLUMB, 1927). According to WENTWORTH (1927), the 300-pound hog yields the same percentage of raw leaf fat but a smaller percentage of prime steam lard when compared to smaller animals; the reason for this lower yield is that an outlet exists for fat backs, jowls and plates, which is more profitable than rendering these cuts into lard. The number of pigs of the lard type in the U.S.A. is rapidly diminishing and breeds such as the Poland-China which formerly furnished this type are now being modified.

In Europe the outstanding example of the lard pig is the Mangaliţa which is usually finished at about 350 lbs. live weight. In view of the recent decline in international trade, Germany has turned to the pig for a supply of fat, not only to avoid imports from the U.S.A., but also as a substitute for vegetable fats obtained chiefly from the East and from West Africa. With this end in view, there has been a tendency towards slaughtering at heavier weights and if the practice proves economically sound, doubtless special breeds will be created or the existing ones will be modified to suit the new requirements.

There is a considerable amount of literature on the relative merits for bacon or pork production of the various breeds and crosses of pigs. Of outstanding importance is the work which has been done by various pig testing stations. Results have been published by many, amongst whom may be mentioned: JESPERSEN and MADSEN (1929, 1931), LARSSON (1928), HANSSON and BENGTSSON (1923—1932), SMITH and CALDER (1928, 1930), CALDER (1931), the Iowa Agricultural Experiment Station (1918, 1922, 1925, 1928a, 1930), LUSH (1931), and FERRIN et alii (1923, 1930, 1931, 1932, 1933).

The methods employed by these stations are fundamentally the same and are briefly summarised under the section „Methods of Improvement". From the results of these pig testing stations, a clear picture can be obtained of the efficiency of food consumption and bacon production of both individuals and breeds. It is clear that such results must be of great aid, not only to the individual breeder whose pigs are tested, but to the general improvement of the pig population. It is, however, more convenient to deal with the problem from the point of view of the animal rather than that of the investigator or the method of investigation. Accordingly, the present section is sub-divided into the following eight divisions:

Meat Qualities.

Type in Relation to Meat Production.

Maturity.

Economy of Live Weight Production.

Slaughter Loss.

Sex in Relation to Meat Production.

Pure Breeds.

Pure Breeds v. Cross Breeds.

It will be noted that within each section the various authorities are usually quoted geographically in the following order: Scandinavia, Europe, United Kingdom, United States, British Dominions, and the rest of the world.

2. *Meat Qualities*

The influence of certain body characters upon ultimate meat production has received attention both in Denmark and in Sweden. JESPERSON and MADSEN(1931) emphasize the importance of selection by length, more especially as this character is not greatly influenced by environment and nutrition, but is a fixed attribute and of great economic value. They obtained a negative correlation between length and thickness of back fat of $r = -0.197$ for males and $r = -0.145$ for gilts, and a positive correlation of $r = +0.2$ between thickness of belly and thickness of back fat. Variations in length and belly were smaller than the variation in the thickness of back fat. LARSSON (1928) found that the thicker the back fat and the belly, the greater the slaughter loss. From Sweden comes the description of the results of SILVERHJELM (1933) in his selection for high quality bacon. With data collected from a herd of Large White pigs, he obtained an increase of first quality bacon from 29 per cent in 1924 to 91 per cent. in 1931. There was also increased efficiency of food consumption and live weight gain.

The results of JESPERSON and MADSEN quoted above were confirmed by RÓŻYCKI (1933) who carried out slaughter tests with the pigs at the Stary-Brześć Pig Testing Station (Poland). He concluded that body length is of the greatest importance in selection for the production of meat for export purposes. Furthermore, he found uniformity·and thickness of the subcutaneous fat to vary greatly, and the percentage of export material to range from 57 to 62 per cent. of the total live weight.

From Germany comes the work of SCHMIDT, VOGEL and ZIMMER-MANN (1929), who investigated the problem of variation and found great individual differences both as regards the quality of meat and the weight of the various parts, as well as in the distribution of fat and flesh. In 1920, SEEDORF (1932) formulated the desirable qualities of the pig and stated that the best pig was one that, together with its offspring, produced the greatest amount of meat and fat with the greatest economy in a specified time. With this aim in view, tests were begun in East Prussia.

That there is great need for improvement in Great Britain may be illustrated by the statement of DUCKHAM (1929a) who estimated that only two-fifths of the carcases handled under the East Anglian Pig Recording Scheme were suitable for the production of good quality Wiltshire bacon. Details of the main faults are of interest. It is stated that „60 per cent. of these carcases were penalized for having too thick back fat, 45 per cent. for uneven fat and heavy shoulders, 30 per cent. were faulted on account of deficient length in relation to weight and 16 per cent. on account of thin flank and belly". The chief reason for faulting was that the time of marketing was too late. Slaughter at a younger age would obviate, to a great extent, the main defects of excess back fat and heavy shoulders. This is confirmed by the fact that 61 per cent. of the carcases of 140—160 lbs. dead weight class and only 10 per cent. of the 180—200 lbs. dead weight class were of good quality. KITCHIN (1931), in a later report, states that 38.6 per cent. of the carcases handled were graded as suitable for first quality bacon. The importance of slaughtering sufficiently early is again emphasized. Under the new Pig Marketing Scheme in Great Britain, breeders have been remarkably quick to appreciate these points and the type of pig now marketed is showing a very considerable improvement.

Under this Pig Marketing Scheme no cognisance is taken of length of back in grading pigs. Only two measurements are made by which the quality of the carcase is assessed, namely, the thickness of the fat over the shoulder and the total thickness of the belly. According to JESPERSEN and MADSEN(1931),there is a positive correlation between the thickness of the back fat and the total thickness of the belly, and this is confirmed by RóżYCKI (1933). Yet the desired British standard is a thin back fat and a thick belly. It is possible that these two points may not be quite so incompatible as at first appears. Thick belly may be one of two types, the thickness being due either to fat or principally to lean meat. If the thickness is due to fat, one would expect a high correlation of the belly with the thickness of the back fat. If the thickness were due to lean, it might be found that a negative correlation existed between back fat and belly.

From pig testing results, CALDER and SMITH (1928) came to the conclusion that bacon produced from Large White pigs was of better quality and of greater popularity with the consumer than that produced by Middle White pigs.

The figures given by SINCLAIR (1932) for the grading of Canadian pigs are of interest, especially when compared with those of Duckham. SINCLAIR states that approximately only 15 per cent. of the hogs marketed in Canada in 1931 reached the top grade of „select bacon" and about 50 per cent. reached the two top grades. Roughly, only half of the pigs reaching the stock yards and packing plants of Canada in 1931 were of a type and quality to process into an article of anything like export standard. In the Province of Ontario, where the bacon type of hog has been established for several years, the two top grades constituted 80 per cent. of the carcases marketed, while in Alberta, where a fixed type has not been established for any length of time, only 34 per cent. of the produce fell into these two classes.

3. *Type in Relation to Meat Production*

Within the past quarter of a century, the type of pig fo-
commercial production throughout the world, has undergone conr

siderable modification. An interesting example of how this has
occurred is to be seen in a series of papers which have appeared from
the United States Agricultural Experiment Stations, especially from
Illinois. It is interesting to note that in the United States, the change
over in the type of pig can be in no small measure attributed to the
professors of Agriculture from the Colleges of the Middle West —
and especially Iowa — who were the first to appreciate the fact that
the type then in favour in the showyard was not the one most desired
by the „packers", and who, by experiments conducted at the college
slaughterhouses, confirmed and elaborated this observation.

The demand of the market for lean meat from young pigs tends to
make selection by type of equal, and even greater, importance than
selection by breed. However, the type produced by a specific cross at
the present time may not be the same as that produced by the same
breeds in the future, and again, the product of a specific cross today
may not suit the market requirements of to-morrow. Many investigations have been undertaken in an attempt to determine the mode
of inheritance of various qualities which go towards the production of
the ideal bacon or pork type of pig, but there are difficulties in such
analysis. Not the least of these is that the type demanded by the
consumer to-day cannot be regarded as final.

HANSSON and BENGTSSON (1930) emphasize the importance of
selecting breeding animals on the basis of conformation, as this seems
to be a reliable indication of good bacon-producing ability. Tests were
carried out in order to ascertain how nearly placing by exterior
judgment agreed with carcase placing. The result of these tests was
as follows: 59 per cent. of the pigs judged by their exterior were
assigned to the price which corresponded to the quality of their
meat; 34 per cent. were placed in a class which differed from the
right place by only 1 degree, while in only 7 per cent. of the cases did
exterior judgment prove to be quite erroneous. (See also JESPERSEN
and MADSEN, p. 000).

LETARD and LEGENDRE (1933) stress the importance of early-
maturing strains within breeds and the need for encouraging the
production of a uniform type of pig of the right age and weight by
means of a grading system.

This essential change in type to suit market requirements is also
emphasized by HAMMOND (1934) who states: „Since, at the present

time, the majority of our pigs are supplying the pork market, and these are of the short, blocky, early-maturing, and quick fattening type, there must needs be a change in type if a pig which has the right proportions at 100 lbs. live weight for pork is to be killed at 200 lbs. live weight for bacon. The type within any breed may be changed by selecting for length and increase in size as was done with the Poland-China breed in U.S.A., when the demand for lard decreased owing to competition with the vegetable fats. Feeding the animal on a lower plane of nutrition with more protein and less carbohydrate food will also help to give a higher proportion of muscle to fat where pigs of the pork type have to be carried through to bacon weights."

The importance attached to selection by type is also emphasized by LUCEY (1931—32) who maintains that the excellence and uniformity of Irish bacon is due to this method of selection being practised for some time. In the production of Irish bacon, farmers are advised to ignore colour and breed and to refrain from using the short thick-set type of pig for breeding. DAVIDSON (1927) has suggested concentrating upon the improvement and efficiency of one type rather than upon several breeds of pigs. Provided the type of pig is good, it should be equally suitable for either bacon or pork.

NORDBY (1932b) discusses type in relation to dressing percentage and concludes that the „chuffy" (pork) type is excessively fat and has a lower quality and quantity of wholesale cuts. LAIBLE, BULL and MITCHELL (1924) and BULL et alii (1929), in their carcase comparisons of „rangy", „intermediate" and „chuffy" types, demonstrated the intermediate type to be superior as regards quality of grade, quantity of bone in the hams, percentage of fat and skin, and general finish. The „intermediate" type was finished and firm at 225 lbs. while some of the „chuffy" type were overfinished and the „rangy" type, unfinished. Carcase measurements showed that the length of carcase, head, neck and legs varied with type, but type had no effect upon depth of chest and length of body proper, as these varied more with the individual rather than with the type. BULL et alii (1929) maintain that type is not a controlling factor in either rate or economy of gain. FARGO and COYNER (1930) state that the type of pig which makes the most rapid gains produces meat of better quality with a higher dressing percentage.

From the results of the Iowa Experiment Station (1918), the more rugged type of pig, such as the Tamworth or Yorkshire, appears to be of greater economic value than the smaller heavier type. The large deeply-covered lard pig was found to be highly efficient. The small closely set refined pig developed early and stayed fat but the cost of production was high; such pigs were found to be liable to pulmonary trouble. In later experiments (1922, 1925, 1928a), very similar results were obtained. The big type of pig makes more rapid gains from weaning until 225 lbs. live weight is reached. The smaller types give a better dressing percentage, but the carcases are too fat. HOGAN (1923) was unable to obtain any significant differences between the various types of pigs, while EVVARD and CULBERTSON (1926) arrived at conclusions similar to those of the Iowa workers. FERRIN (1933) states that the work at Minnesota shows depth and width of body to be pronounced in the early-maturing type of swine.

SINCLAIR and SACKVILLE (1927) consider that breeding plays a greater part in the production of the bacon type of pig than feeding. Another paper which draws attention to the importance of selection by type is one from New Zealand (1930).

Meat qualities such as thickness of back fat, texture of fat, „marbling" of the lean with the fat, and the percentage of dead to live weight, are undoubtedly influenced to a considerable extent by disease, husbandry and nutrition. MURRAY (1934) found that depth of side increases with rate of gain and that back fat increases with increase in the rate of gain. To judge from the Danish, American and German work quoted above, these qualities are also appreciably conditioned by genetical factors.

4. Maturity

Type merges into the whole question of early maturity, which KRONACHER (1930) attributes principally to genetic factors; this point of view is supported by the reports of the testing stations in Denmark, Sweden, the United States of America and Scotland.

LARSSON (1928) deals with the type of animal and the degree of fattening and slaughter loss. He gathered his material from various pig testing stations in Denmark. The thickness of back fat was found to be the best measure of the degree of fattening and bacon quality. A negative correlation was obtained between degree of fattening and body length, the longest animals being the least fat. A weak positive correlation was found to exist between thickness of back fat and thickness of belly. In conclusion, LARSSON states that the large type of animal is the best bacon producer, as it reaches the desired weight of 90 Kgs. at a younger age and with less fat formation. LUND, BECK and ROSTING (1925), in their comparison of the Danish Landrace and the Yorkshire breeds, found that the Landrace attained slaughter weight 10 days earlier than the Yorkshire.

The observations of SCHMIDT, VOGEL and ZIMMERMANN (1929) concerning the variations in type of the Landschwein, lead to the conclusion that no correlation exists in respect of the development and food consumption between animals of the two particular types which occur in the Landschwein breed.

According to MOSKOVITS (1931), the capacity for flesh formation is more dependent upon the hereditary disposition than is that of fat formation. As flesh forms only during the period of growth, too early maturity results in over-fattening, whilst on the other hand, too late maturity results in over-aged animals. Great importance is attached to the value of crossing, as type is more important than breed. The aim is uniformity which must ultimately eliminate breed differences.

At the Scottish Station, SMITH and CALDER (1930) found the average age at which the pigs reached 200 lbs. live weight to be 229 days, but the different litter groups showed considerable variation.

The following variations in the rate of maturity shown by litters produced by one sow and sired by four different boars are reported by BUCHANAN SMITH (1934):

By boar   I a litter of   9 made 0.83 lb. live weight gain per day.
  „   „   II „   „   „ 10   „   0.84 „   „   „   „   „   „
  „   „   III „   „   „   9   „   1.05 „   „   „   „   „   „
  „   „   IV „   „   „ 10   „   1.05 „   „   „   „   „   „

The first two litters required about 240 days to reach bacon weight, the third 200 days and the fourth only 195 days.

In America, CULBERTSON, EVVARD, KILDEE and HELSER (1931) reporting on the Iowa Station Experiments, show that in reaching 225 lbs. live weight, the litter making the quickest gain was 22 per cent. faster than the average of 44 litters tested. This litter variation was also commented upon by FERRIN, ANDERSON and JOHNSON

(1932), who pointed out that in this variation lies the opportunity for selection and development of strains of pigs which are highly efficient. FARGO and COYNER (1930) found that their fastest growing litter gained at the rate of 1.86 lbs. and the slowest, of 1.15 lbs. per day. Furthermore, the quicker growing pigs consumed less food (360 lbs. against 485 lbs.) and produced carcases of top quality with a dressing percentage above the average.

RóYŻCKI (1933) calculated the range of variation in weight within tested litters to be 3.5—20.0 per cent.

Another factor to be considered, and one which is closely allied to maturity and live weight gain, is birth weight. Opinions are diverse as to the effect of birth weight upon the final weight. AXELSSON (1928) obtains correlations of $r = + 0.6765$ and $+ 0.7083$ between the litter size at birth and weight at 3 weeks old. WILD (1927), on the other hand, states that the birth weight is no indication of the subsequent gain in weight. STAHL (1930) states that, as would be expected, litters of more than 13 or 14 do not give such satisfactory results from the point of view of ultimate production. STAHL maintains that there is a close connection between the weight of the breeding sow and that of the piglings. Another table which is of interest in that it shows that litter size does not appear materially to affect maturity (expressed as weight at eight weeks old) is that from New Zealand, (1930):

Number of pigs per Litter	Number of Litters	Average Litter Weight	Average Weight per Piglet
		lbs.	*lbs.*
5	44	151	30.2
6	69	185	30.8
7	85	206	29.5
8	86	238	29.8
9	58	252	28.0
10	41	276	27.6
11	13	319	29.0
12	3	361	30.0

HAMMOND (1922) has investigated the rate of maturity of different breeds of pigs, and his results are summarised in Table IX of his paper: —

Rate of Maturity of different breeds of pigs
Weight as percentage of weight at 11 months old

	Pen of 2 classes				Single pig classes
Age in months	3	5	7	9	9
Middle White	20	31	55	85	86
Berkshire	21	29	54	84	81
Large White	18	22	—	81	81
Tamworth	—	—	72	82	86
Large Black	—	—	65	79	—
Somersetshire.	—	32	—	80	—
Lincolnshire Curly Coated . .	14	—	67	79	96
Dorset	—	—	—	—	68
Small White	—	—	55	79	—

5. *Economy of Live Weight Gain*

That certain individuals and breeds are better „feeders" than others has long been recognised by the practical stockman. In the pig it has been clearly shown that some animals utilise their food more economically than others. In the cost of production of a bacon pig, it is reckoned that the food costs constitute approximately 75 per cent. of the whole. Thus, any economy in this direction has far-reaching results. The reports of the pig testing stations are particularly valuable for genetic enquiry, in that the nutrition and environment at any one station are held as stable as practicable and hence the genetic aspects stand out in sharp relief. It is not possible to compare exactly with each other the reports of the various testing stations, though the Danish ones are so well controlled that the results of one may, within reason, be compared with those of others in Denmark and Sweden. Stations outside Scandinavia have different methods and different finishing weights.

Some interesting figures are given by BECK (1931). The food consumption per pound of live weight gain varies from 2.93 lbs. to 4.26 lbs. per litter group; the average, which varies slightly but not significantly with the breed, is about 3.25 lbs.

SCHMIDT, VOGEL and ZIMMERMANN (1929) emphasize the value of pig recording for disclosing the most productive strains. They found

that the energy value of food required by different pigs under the
same conditions varied from 250 Kgs. to 336 Kgs. The digestible
protein also varied from 33 Kgs. to 43 Kgs. Similarly, the results of
WELLMANN (1930) indicated that while Mangaliţa pigs required 9.2
Kgs. of food per 1 Kg. live weight increase, Berkshire pigs needed
only 8.1 Kgs. to produce an equal gain.

BUCHANAN SMITH & CALDER (1930) also discuss this question.
Their data comprised 35 litter groups from the Scottish Pig Testing
Station. The average food consumption per lb. of live weight gain
was found to be 4.56 lbs., the highest 5.51 lbs., and the lowest 3.68
lbs. The early-maturing pigs proved to be the more economical
producers. A certain variation was due to the season of the year.
BUCHANAN SMITH (1934) reports the following average food requirements for litters sired by four different boars:

Boar   I   3.9   lbs. meal per lb. of live weight gain.
  „   II   3.13 „   „   „   „   „   „   „   „
  „   III 3.9   „   „   „   „   „   „   „   „
  „   IV   3.31 „   „   „   „   „   „   „   „

The whole question of growth in pigs is interestingly discussed,
though hardly from a genetical aspect, by HAMMOND (1922).
Particular reference might be made to the note on the change in
growth rate of the Berkshire pig, the results of which are to be found
in Table XVIII of that paper (see page 96 *infra*). Results of the trial
carried out at the Lord WANDSWORTH Institution (Anon. 1923),
although based on very small numbers, indicate that while pure-
bred Large Black pigs required 3.3 lbs. of meal, cross-breds required
only 3.1 lbs. for every pound of live weight increase.

CULBERTSON and EVVARD (1925) observed that pure-bred Poland-
China piglings (high grade Poland-China sow × pure-bred Poland-
China boar) made greatest average daily gain and required less food
as compared with half and three-quarter bred Poland-China ×
European Wild.

Later, CULBERTSON, EVVARD, KILDEE and HELSER (1931) analysed
the proportions of the food eaten by the various litter groups tested
under the National Swine Record of Performance Scheme. All litters
tested in this scheme are fed standardised rations consisting of shelled
corn, supplemental mixture (protein) and mineral mixture, until each

group attains an average live weight of 225 lbs. These authors point out that when the litters are arranged in order of merit according to the amount of food consumed per 100 lbs. live weight increase, the litter ranking first is not necessarily the one which has made the most economical gain since it may have consumed relatively more of the supplemental mixture — the expensive portion of the ration — than other litters showing a higher total food consumption. For example, out of 44 litters tested by these workers, the one placed first, with a total food requirement of 358 lbs. per 100 lbs. live weight increase, consumed 71 lbs. of the supplemental mixture, while the litters placed second, third and fifth, with total food requirements of 361 lbs., 363 lbs. and 368 lbs., required 55 lbs., 60 lbs. and 58 lbs. respectively of supplemental mixture. Thus those three litters showed more economical gains than the first. The average amount of food consumed per lb. of live weight gain for 44 litters ranged from 3.58 lbs. to 4.68 lbs., while the average for all lots was 3.97 lbs. CULBERTSON, KILDEE, HELSER and HAMMOND (1932, 1933), in litter comparisons under the same scheme, obtained a range of from 346 to 451 lbs. of food required per 100 lbs. live weight increase, with an average requirement of 396 lbs., for 20 litters tested during 1931, and a range of from 344 to 413 lbs., with an average of 376 lbs., for 17 litters tested during 1932. Similar variations were obtained in the amount of supplemental (protein) mixture consumed.

Similarly, FERRIN, ANDERSON and JOHNSON (1931, 1932) obtained, per 100 lbs. live weight increase, ranges of 342—414 lbs., 345—409 lbs., and 331—384 lbs. for 9, 12 and 13 litters tested in 1929, 1930 and 1931 respectively. Their figures also show that litters which consume the smallest total amount of food per 100 lbs. live weight increase do not necessarily make the most economical gains, owing to a relatively high consumption of the expensive protein constituents of the ration.

MOHLER (1933), chief of the Bureau of Animal Industry, United States Department of Agriculture, reports that twelve litters of piglings born in the spring of 1932 were tested against seven litters born in the autumn of the same year. „Four pigs were used to represent each litter. They were on test from 72 days of age until they reached a final weight of approximately 225 lbs. The variation in average daily gain for the twelve groups of spring pigs was from 0.56 to 1.41

pounds. Feed consumed per 100 pounds gain ranged from 360.7 to 503.6 pounds. The seven groups of fall pigs ranged in average daily gain from 0.98 to 1.63 pounds, whereas the feed consumed per 100 pounds gain ranged from 336.2 to 396.1 pounds. Wide differences in efficiency of production were associated with differences in breeding among the groups. The importance of basing selections of breeding animals on progeny tests is emphasised by these results." MOHLER also gives some interesting figures concerning the influence of the age of the dam. Pigs were divided into three groups according as to whether their dams were gilts from gilts, gilts from old sows, or old sows; the three groups were fed under uniform conditions to a market weight of approximately 200 pounds. „During this feeding period the pigs whose dams were gilts from gilts made an average daily gain of 1.35 pounds, those from gilts from old sows 1.43 pounds, and those from old sows 1.40 pounds. The respective quantities of feed consumed by the pigs per 100 pounds gain were 382.8, 399.2 and 416.7 pounds."

BAIRD (1923) and MILNE (1920), of Canada, both investigated this problem of economy of live weight gain. The former obtained greater economy of gain with the Yorkshire than with the Berkshire, whilst the results of the latter showed no significant differences between Berkshires, Yorkshires and Duroc-Jerseys. In the experiments of ROTHWELL (1924), 15 Yorkshire pigs made an average daily gain of 0.69 lbs. per head during 42 days, as compared with a gain of 0.79 lbs. made by the Berkshires.

The figures given above amply prove that there is genetic variability underlying the economy of live weight gain. This gain may be considered as comprising two distinct aspects which interact, viz. early maturity and the ratio of food consumption to live weight gain. It has been shown (DUNLOP, 1933) that other factors, namely, age, sex, condition, and previous growth rate, have no effect on the rate of live weight increase of pigs.

6. Slaughter Loss

Of great importance is the proportion of the finished carcase which is available for human consumption. Work on the killing percentages has been carried out in connection with pig testing and economy of live weight gain investigations.LARSSON (1928) found that the thicker the back fat and belly, the greater the loss during slaughter. HANS-SON and BENGTSSON, in their many investigations of the comparative values of the Landrace and Yorkshire, are generally in agreement as to the slight superiority of the Yorkshire regarding the loss during slaughter, the slaughter loss percentage being approximately 26 and 25 respectively.

RÓŻYCKI (1933) found that the percentage of export material ranged from 57 to 62 per cent. of the total live weight gain. This figure is not the same as live to dead weight percentage: but is an indication of the variation which occurs.

SCHMIDT and VOGEL (1928) calculated that the slaughter loss averaged 21 per cent. of the live weight and was approximately equal for both heavy and light animals. There was, however, great individual variation, ranging from 25.69 per cent. to 16.85 per cent. In a later paper (1930a), the slaughter loss for male and for female was found to be about equal, although the total dressed weight of the male was somewhat higher.

HAMMOND (1922) has investigated this problem in great detail. His table showing the percentage increase in carcase weight with advance in age is most interesting. This increase is due to the development with age of muscle and fat without a corresponding increase in the size and weight of other organs of the body. He tabulates these facts as follows:

Changes in the proportional development of Berkshire pigs from Period I (1903—6) to Period II (1907—13)
Percentage of live weight

	3 months			5 months			7 months			9 months			11 months		
Period	Carcase	Pluck	Intestines, etc.	Carcase	Pluck	Intestines, etc.	Carcase	Pluck	Intestines, etc.	Carcase	Pluck	Intestines, etc.	Carcase	Pluck	Intestines, etc.
I	67.4	6.5	26.1	78.2	5.3	16.5	82.1	4.7	13.2	82.7	4.5	12.8	83.7	4.5	11.8
II	77.3	5.3	17.4	77.6	5.5	16.9	80.7	4.8	14.5	83.1	4.5	12.4	82.8	4.3	12.9
	Numbers from which calculated														
I	1			23			9			18			11		
II	26			35			23			51			27		

HAMMOND mentions the following three main factors which he considers to be responsible for the variation in carcase percentage at any one age: 1) differences in fatness, 2) early maturity, and 3) the amount of food in the alimentary canal. He also summarises the work of various dieticians which indicates that various types of rations, rickets, starvation, etc. will ultimately affect the proportions of the body and therefore influence the slaughter percentage.

HENNING & STOUT (1932) who investigated the factors which influence the dressing percentage of hogs, could obtain no significant effect of production factors upon yield. Pigs receiving corn alone gave as good dressing percentages as those receiving more expensive and better balanced rations. Furthermore, they could find no difference in yield between pure-bred or cross-bred pigs; age also appeared to have no effect. The four factors positively correlated with yields are: the amount of lard, the corn-hog ratio, sex, and live weight. Their results are shown in the following table:

The Correlation of Yield of Hogs, slaughtered under Federal Inspection, with the Percentage of Lard, Percentage of Barrows slaughtered, Live Weight, and the Corn-hog Ratio for the United States for the years 1923—1930
(Check sum method used)

Yield correlated with	Unadjusted coefficient of correlation	Coefficient of determination
		Pct.
Lard	0.621	38.62
Corn-hog Ratio , ...	0.478	22.84
Barrows	0.407	16.56
Live weight	0.229	5.24

It can be seen that the amount of lard (or degree of fatness) has the greatest effect upon the loss at slaughter.

In the investigation of NORDBY (1932b), no significant difference was detected between the killing percentage of a group of five pigs of the „chuffy" type and a similar group of the large type. BULL and LONGWELL (1929), on data gathered from 189 pigs slaughtered at

175, 225 and 275 lbs. respectively, found that the very „chuffy" type of pig dressed higher when slaughtered at 175 lbs. than the intermediate and rangy types, but at other weights there were no significant differences in the dressing percentages. CULBERTSON et alii (1928) obtained dressing percentages ranging from 80.4 to 82.4 per cent. from data which included 10 litters fed and slaughtered in an identical manner.

7. Sex in Relation to Meat Qualities

The general opinion appears to be in favour of the superiority of the female as an economical producer of first class meat. DUCKHAM (1929a), from data gathered by the East Anglian Pig Recording Scheme, finds that gilts tend to produce bacon of superior quality. On the whole, gilts are seldom faulted for insufficient length in relation to weight, over-thick back fat, heavy shoulders, or thin flank, all of which are very important considerations in the economic production of bacon. PARK (1930) and BULL and OLSON (1931) are all in agreement as to the superiority of hams produced by gilts. LACY (1932) found that gilts yield a definitely higher proportion of the more desirable cuts, but with a slightly lower quality of bacon. On the other hand, YATES (1934), in the course of a feeding experiment, found no trace of any sex difference in performance between pigs of opposite sex of the same litter and receiving identical treatment.

The dressing percentage of barrows and gilts is approximately equal, the difference being only 0.3 per cent. For example, results of Canadian investigations fail to show a significant difference in the dressed yields of barrows and sows. The differences were too small and inconsistent to be of value from a practical point of view.

AXELSSON (1933) analysed various body characters in both the male and female and found that in all cases the female was superior as regards maturity and the rate of development of desirable meat characters. He is, therefore, strongly in favour of females for bacon production. BECK (1933) also found that in all breeds, and especially in the Landrace, sows (i.e. gilts) were superior to boars. The slaughter loss amongst males was 5.5 per cent. and among females only 3.1 per cent.

SCHMIDT, VOGEL and ZIMMERMANN (quoted by KRALLINGER, 1930) are in favour of the castrated male for purposes of meat production. REED (1922) was able to produce a more rapid gain in males than females, while DAVIDSON (1927) discusses the possibility of maintaining a bacon type of pig for both pork and bacon production and suggests that this might be done successfully by using hogs for pork and gilts for bacon.

8. *Pure Breeds*

Several of the principal breeds have been the subject of investigation. Of these investigations, greatest attention must be directed towards the work carried out in Scandinavia in breeding bacon pigs for the English market. HANSSON and BENGTSSON, in their numerous papers, have reported on the relative merits of the two breeds which the Swedish farmers have found by experience to be the best suited both to their own methods of production and to the requirements of the English producer, viz. the Swedish Landrace and the English Large White or Yorkshire. Their data were collected from performance tests carried out at the Åstorp Pig-Testing Station. For these tests, 4 piglings, 2 of each sex, from each tested litter were sent to the Station. During growing and fattening periods the usual records were kept.

These writers, (1928), state that the Landrace gave a somewhat smaller yield of bacon for export; the Yorkshire was superior in firmness of back fat, ham development and belly bacon, whilst the Landrace had greater body length, thickness, and distribution of back fat. When slaughtered, 67 per cent. of the Landrace pigs were placed in Class I, while the corresponding figure for the Yorkshires was not more than 52 per cent. They emphasize the fact that the production of first class bacon for export does not necessarily mean less rapid growth or greater food consumption. In the following year, (HANSSON and BENGTSSON, 1929a), the results of tests of 63 experimental lots, 12 of the Landrace and 51 of the Yorkshire, during the year 1928 are summarised as follows:

Comparison of the relative values of the Landrace and Yorkshire

	Landrace	Yorkshire
Gain per animal per day in gms	636	619
Food consumption per kg. gain (food units)	3.56	3.73
Bacon, %	62.2	62.3
Wastage, %	11.8	12.5
Slaughter loss, %	26.0	25.2
Grading		
No. of pigs in Class I, %.	53.2	55.0
No. of pigs in Class II, %	29.0	25.4
No. of pigs in Class III, %.	17.8	19.6
Economic value (No. of points)	74.0	72.0

The period covered by these figures is from 8—9 weeks old to the attainment of an average live weight of 90 kgs. From the above table it can be seen that the difference between the two breeds is negligible. In a later paper (HANSSON and BENGTSSON, 1931), the more important results of tests of 106 experiment lots, 26 of the Landrace and 80 of the Yorkshire, during the year 1930, are summarised as follows:

	Landrace	Yorkshire
Age in days at slaughter	169	176
Gain in weight per pig per day (gms.) .	689	652
Number of food units per Kg. gain . . .	3.40	3.57
Export bacon (per cent.)	60.9	61.2
Slaughter waste (per cent.)	14.2	14.2
Slaughter loss (per cent.)	24.9	24.6
Animals in Class I (per cent.)	38	47
,, ,, ,, II (,, ,,)	33	27
,, ,, ,, III (,, ,,)	29	26
Economic value (No. of points)	75.5	73.8

The tests began when the average body weight was about 20 Kgs.

and the animals were slaughtered at 88—89 kgs. It is pointed out that, in the above figures, kidney fat, formerly included under bacon, is counted as slaughter waste; consequently the proportion of bacon is somewhat lower compared with earlier results. It will be seen that the averages for both breeds are fairly close; the authors emphasize, however, that considerable differences occur in some of the groups.

LUND, BECK & ROSTING (1925), in a similar comparison of the Danish Landrace and the Yorkshire breeds, had very much the same results. Here the Yorkshire showed slight superiority as a producer both of good quality and of quantity of export bacon, while the Landrace had a somewhat longer body and attained slaughter weight 10 days earlier.

SCHMIDT, VOGEL & ZIMMERMANN (1929), in a comparison of the slaughter loss of the Improved Landschwein and the White Edel-schwein, found this to vary between 18% and 21% for both breeds. They also found no difference in the distribution of fat and flesh and the weight of separate parts. Their observations have led them to conclude that there are no distinct breed differences in respect of performance characters such as development and food utilisation.

As stated before, WELLMANN (1930), working in Rumania with the Berkshire and Mangaliţa breeds, shows that the former require 8.1 Kgs. of food per 1 Kg. live weight increase, and the latter 9.2 Kgs. CALINESCU (1929) discussing the relative yields from Mangaliţa and Berkshire pigs, gives these figures — Mangaliţa yield 54% fat and Berkshire 37% fat. The work of RACZ (1931) is of interest in that it shows the importance of continued selection within the breed if steady improvement is to be achieved. He demonstrates in the following table that selection has been responsible for great improvement in the Mangaliţa breed:

	1928	1930
Weight at birth of Mangaliţa litters .	8.66 Kgs.	8.89 Kgs.
„ „ 28 days „ „ .	30.4 „	31.3 „
„ „ 56 „ „ „ .	55.4 „	58.2 „
„ „ weaning „ „ .	60.7 „	68.8 „

In Canada, BAIRD (1923) carried out feeding tests with Yorkshires and Berkshires and found the latter breed to give smaller gains with greater costs. MILNE (1920) obtained the following figures for food consumption per lb. of gain: Berkshires 5.15 lbs.; Yorkshires 5.15 lbs., and Duroc-Jerseys 5.71 lbs. ROTHWELL (1924) obtained results in favour of Berkshires rather than Yorkshires. As a result of five years of testing at the Ontario Agricultural College, the following breeds were placed in this order of merit: Duroc-Jersey, Berkshire, Tamworth and Yorkshire.

The table of REED (1922), given below, is of interest for breed comparison. It can be seen that, on the whole, the Yorkshire made the most economical gains; it also produced the type of carcase most nearly approaching that required by exporters.

	Live Weight gain per Day		Average food cost of gain per lb.
	Male, lbs.	Female, lbs.	
Yorkshire	1.123	0.997	3.06 cents.
Berkshire	1.089	0.815	3.25 „
Duroc-Jersey	1.303	0.930	3.40 „

Results from New Zealand (1930) appear to indicate the superiority of Large Black and Large White as bacon producing breeds.

It should be noted that in all breed comparisons due allowance must be made for the large differences which are likely to occur between the average results of the progeny of different boars of the same breed, as pointed out by MURRAY (1934).

9. *Cross Breeding versus Pure Breeding*

For commercial meat production, many producers use the first cross between breeds, the backcross, or the progeny of the first cross by a sire of a different pure breed. There are two reasons for the undoubted success, in many instances, of one of these policies. Firstly, from the genetical point of view, there is reason to believe that crosses between certain breeds exhibit the phenomenon of heterosis. Secondly, from the practical point of view, it is sometimes advantageous that the dam of a first cross be of a hardy breed which can withstand climatic conditions and rear her young under a somewhat rigorous environment, while the sire comes from a more quickly maturing and somewhat less hardy breed. The progeny can then be raised more economically than if they belong to the breed of the sire, and they will also reach maturity more quickly than if they belong to the breed of their dam. Furthermore, there is evidence which indicates that in certain breeds the maternal instincts are more highly developed than in others and therefore one possible justification for this practice is that the progeny, while inheriting the meat qualities of their sire, benefit also from the mothering of their dam. However, in pigs, the second policy holds to a much lesser extent than in cattle or sheep, except in the case of the grazing or „pasture" pig. As a rule, bacon production is more intensive than beef production, while mutton is the least intensive of all. Sheep breeders, therefore, are the most likely to adopt cross breeding. Let us now examine the data on the pig.

AXELSSON (1929a) obtained the greatest gain with the pure-bred Swedish Large White, and does not favour cross breeding if the greatest economy of gain is the object. On the other hand, BECK (1931) states that there is no real difference between the gains made by the Yorkshire, the Danish Landrace, or their first crosses. Apparently he did not find evidence in favour of either cross or pure breeding. In support of selection within the breed rather than crossing as a means of improvement may be mentioned HANSSON & BENGTSSON (1923 etc.). They state that excellent results can be obtained by selection of families within a certain breed, as considerable variation occurs in individual strains.

RICHTER, HEMPEL, OHLIGMACHER and RODEWALD (1928), taking measurements at the fifth and the tenth week, found the cross-breds to have greater depth, larger chest measurements and altogether better conformation than the pure-breds, which had less width, larger heads and sloping pelves, thus giving the impression of late maturity. SCHMIDT and co-workers (1926, 1929) are also in favour of cross-breds for high efficiency of production and good utilisation of food. They (1929) draw attention to the occurrence of great individual variation both in respect of fat and flesh distribution.

RACZ (1931) found that at weaning, cross-bred Mangaliţa × Berkshire weighed 4 to 5 Kgs. more than pure-bred Mangaliţa, and at 1 year the cross-breds weighed 75 Kgs. and the pure-breds only 60 Kgs. In other words, the cross-breds weighed 10% more at weaning and 15% more at 1 year than pure-bred Mangaliţas. On the other hand, the pure-bred animals gave a better distribution of fat and lean than the cross-bred. KRONACHER (1930), who has probably a wider experience than any other investigator, reports that he has never observed the phenomenon of heterosis in his many crosses. KENNESSEY (1930) carried out a biometric comparison of Berkshire pigs which had been raised in Hungary, with those imported from England, and found the former to be in no way inferior. He states that Berkshires are much used for crossing with the Mangaliţa to produce a type of pig excellent for fattening.

IVANOV (1933) describes the results obtained by crossing Large Whites bred in Ukraine with the local native breed. The cross-breds showed greater fertility and a larger number of live offspring per litter, the average litter size being 13.9 and 12.9 respectively. The total number of all offspring from 100 sows (including sterile ones) showed an excess of 15 per cent. in favour of cross-breds. Gain in weight and percentage of mortality were also in favour of cross-breds

HAMMOND (1922) expressed the opinion that crossing increases vigour, size of offspring and early maturity. From his data, which were drawn from the Large, Middle and Small White, Large and Small Black, Berkshire, Tamworth, Lincoln Curly Coat, Dorset and Somerset breeds, he concludes that there exists a breed difference in the ability to utilise food rather than actual food capacity. There is also a breed difference in the time taken to attain maximum development, the Middle White being the earliest to mature. Cross-breds were usually found to be heavier than their parents. Later, BUCHANAN SMITH and CALDER (1930) obtained no significant difference between the live weight gain of 15 pure-bred Large White litter groups and 13 groups of Large White × Large Black.

Work carried out at the Lord WANDSWORTH Institution, (Anon., 1923) gave results in favour of the cross-bred pig. The trial, which consisted of only 9 pure Large Black and 9 Berkshire × Large Black cross-bred animals, proved that the cross-breds were superior both as regards meal consumption and live weight gain. GRANDI (1931) is also in favour of cross-bred stock for the production of bacon or pork pigs. The crosses found to be most satisfactory were, for bacon, Large White male × Large Black female, and for pork, Large White × Middle White cross-bred female by a Middle White boar. Pure-bred Middle Whites were found to be somewhat blocky.

According to LUCEY (1931—32), the marked improvement in the quality and uniformity of the Irish pig population is due to rigid selection and to the use of premium boars which have been chosen according to the length and depth of side, fineness of shoulder, and ham development. Again the value of the pure-bred animal selected for type is emphasized.

This much discussed question as to the relative merits of cross-bred and pure-bred pigs has received a great deal of attention from workers in the United States. COOPER (1914) is in favour of pure-bred animals as the most economical producers. LUSH (1923) crossed Duroc-Jerseys and Berkshires and obtained more satisfactory results from cross-breds than from the average of the two parents. He points out that while the F_1 are likely to give good results, the F_2 are usually slightly inferior. For this reason he concludes that it is safer to limit crossing to distinct types within a breed. HICKS (1922) gained 4.75 $ profit from cross-breds (Yorkshire × Duroc-Jersey) and only 3.68 $ profit from pure-breds (Yorkshire). CRAFT (1927—30) states that out-bred pigs make a greater gain in 8 per cent less time than pure-bred pigs.

GRIMES and SEWALL (1928) state that, as the grade of the pig increases, so the food consumption decreases owing to improved food utilisation: 87.5 per cent of pure-bred pigs reached the final stage 61 days earlier and required 167 lbs. less food for each 100 lbs. of gain than did scrub animals. Pigs that were 50 per cent pure-bred gave a more satisfactory performance than those that were 75 per cent pure-bred. This work was continued (1929, 1930) and a pure-bred boar was used for 3 generations with marked improvement in type, quality, and ability to make cheap and rapid gains. This may be seen in the following table:

	Time required to reach 200 lb. live weight	Average daily gain	Food required per 100 lb. gain
Scrub pigs . .	244 days	0.95 lb.	465.35 lb.
50 % grades .	201	1.18	403.37
75 % „ .	201	1.19	387.63
87.5% „ .	187	1.26	381.52

TINLINE (1922, 1923), in an attempt to determine the best cross, obtained the most rapid gains from Yorkshire boar × Berkshire sow crosses. He found Duroc-Jersey × Yorkshire to give the most rapid and economical gains, but the carcase type was not so desirable for bacon. HAYWARD, (as quoted by KRONACHER 1924) is also in agreement as to the superiority of cross-breds with regard to size, growth and vigour. CULBERTSON and EVVARD (1925), at the Iowa Station, compared the average daily gain made by pure-bred Poland-China and cross-bred Poland-China × European Wild piglings, and, as might be expected, found the former more satisfactory.

ROBERTS and LAIBLE (1925) carried out double-mating experiments in which Duroc-Jersey sows were mated first to a Poland-China boar and then to a Duroc-Jersey boar. Six of the piglings were Duroc-Jerseys and four were cross-breds. The pure-bred averaged at birth 3.25 lbs. and the cross-breds 3.75 lbs. Two of the former raised to 6 months averaged 185.5 lbs., while the four cross-breds averaged 235.2 lbs. These differences are attributed to heterosis, but the results were obtained from only one litter and the subsequent work of CARROLL and ROBERTS must be considered. These workers (Illinois Experiment Station, 1926) double-mated Duroc-Jersey and Poland-China boars with 5 Duroc-Jersey and 5 Poland-China gilts. The cross-bred and pure-bred offspring showed no significant difference in vigour or size at birth, and no marked superiority was exhibited by either group in the utilisation of food and economic gain. Later, (1928) CARROLL and ROBERTS, found no indication of significant differences in the rate of economy of gain made by cross-bred and pure-bred pigs farrowed in the same litter.

Similarly at Iowa, SHEARER et alii (1926) also practised double mating as a means of determining the difference in growth etc. between cross-bred and pure-bred pigs. They double-mated pure-bred Poland-China sows with Duroc-Jersey and Poland-China boars and obtained 5 litters in which there were 31 cross-breds and 21 pure-breds. At birth, the pure-breds were slightly heavier, averaging 2.89 lbs. as compared with 2.66 lbs., but only 67 per cent. of the pure-breds were classified as excellent against 71 per cent. of the cross-breds. Again, 9.5 per cent. of the pure-breds and only 3.2 per cent. of the cross-breds were born dead. The rate of increase in live weight up to weaning was similar for both groups. For the period from weaning to 225 lbs. live weight, the cross-breds showed their superiority both as regards rate of gain, capacity and utilisation of food, body growth as determined by measurements, and calculated profit over food cost.

Both these investigations at Illinois and Iowa were conducted by the most competent workers. The difference in the results obtained from these two apparently similar experiments is probably due to the strains of the parent breeds. In Iowa they „nicked"; in Illinois the breeds, while not antagonistic, probably possessed genetic constitutions of greater dissimilarity. This emphasises the point that strain within a breed is frequently as important as the breed itself.

CRAFT (1924—26) studied the effect of limited inbreeding and outcrossing in swine. At birth, the outcrosses showed a slightly heavier average weight but the differences were not significant. At weaning, the limited inbreeding groups averaged 5.2 lbs. heavier but there were no other significantly different characteristics. Surplus animals were fattened to market at 200 lbs. live weight but again results for each group were similar. M'PHEE (1932), who found that crosses of inbred Tamworth × Chester-White gave uniform and vigorous F_1 stock, supports the theory that heterosis is responsible for the superiority of certain first crosses over pure-bred animals.

REED (1922) was able to produce more rapid gains with the Berkshire × Duroc-Jersey cross than with the Yorkshire × Berkshire cross. In the feeding tests of TOOLE and KNOX (1926), pure-bred Yorkshires, Tamworths and Berkshires gave more satisfactory results regarding consumption than the first crosses of those breeds. Also from Canada comes the account of a double-mating experiment. Duroc-Jersey sows were mated both to a Duroc-Jersey and to a Yorkshire boar. The result was nine cross-bred and nine pure-bred piglings. At 180 days of age the cross-breds averaged 16.5 lbs. heavier than the pure-breds.

For economical production in China, HABU (1932) recommends crosses of Poland-China × Berkshire (both ways), Berkshire × Large White, and Middle White × Poland-China.

On the whole, it would appear that the first cross between certain breeds is desirable for commercial meat production, and particularly for bacon pigs. The fact that, in certain instances, the cross-bred is not superior to the pure-bred is an indication that hybrid vigour is not a universal phenomenon but depends both upon the breeds themselves and the strains within the breeds. This is precisely what the science of genetics would lead us to expect. In view of the fact that the different breeds of pigs are as a rule bred less for their suitability for a particular environment (as is the case with sheep) than for particular qualities, this practice of cross-breeding may be taken as evidence of the existence of heterosis.

XI. METHODS OF IMPROVEMENT

1. *Introduction*

It is safe to say that the full benefit of genetics will not be obtained until some standards of production are established. Before the scientist can control the transmission of heritable characters and pass this information on to the practical breeder, he must have a precise definition of these characters. It is natural, therefore, that there should be a desire to establish a standard for the pig whereby this animal may be improved as an economical producer of food. This desire has already found expression in many parts of the world. It is the object of the present section to discuss what has been achieved along these lines in the different countries. Both Governments and breed societies are considering, or have already adopted, methods either of advanced registry or of more detailed pedigree registration; hence it would seem opportune to make a critical survey of such work as has been accomplished and to direct attention to those methods which experience indicates to be best adapted to meet the needs of the present day.

Any system of more elaborate registration than the mere record of ancestry is a perfectly logical step in the evolution of methods of animal improvement. No one will deny the tremendous impetus given to the British breeds of livestock by the establishment of herd books, an impetus which is still felt in our market for the export of pedigree stock. In the early days of the various breeds, the breeders of pedigree stock were not as numerous as they are now, and there was more personal intercourse amongst them; it is probable that each breeder was intimately acquainted with the quality of the stock of most of his competitors. To-day, a larger number of pedigree breeders are engaged in the business; they are distributed over a much wider area and therefore much of this intimate knowledge of the type of stock in relation to pedigree is no longer available. Further, in the early days of pedigree breeding, every local market had its peculiarities and the local breed was bred to suit that need. To-day, local peculiarities in demand tend to be submerged into the popular standards demanded by the general consumer.

Under these circumstances, it is not surprising that there has arisen a certain amount of discussion as to whether pedigree is really of much value. This criticism is not without some justification. The time is now ripe to take steps to increase the value of pedigree by recording (not necessarily in the herd books) the productivity of the pig as well as the fact that it has been bred pure. A pig with a pedigree implies that it is aiming at something higher than the average. A pig with a record provides information as to whether it has or has not achieved this.

Thus, any proper system of advanced registry in pigs will really be an accumulation of facts concerning the growth of pigs in general and a certain individual hog in particular. In effect, it becomes the science of the type of the pig. To get this knowledge or science we must employ tools upon which, as Lord KELVIN once declared, all accurate scientific work must be founded, namely weighing and measurement. Thus, improvement of the breed or race is closely associated with the advancement of our knowledge concerning the genetic qualities of the pig, a fact well illustrated in the preceding section of this paper. That these two points are so closely connected furnishes additional reason why a further section of the paper can, with profit, be devoted to the practical methods employed in different parts of the world for the improvement of the pig.

2. *Litter Testing*

The Litter Testing Stations which are situated in many countries in Europe and America are of the greatest importance to the pig industry. They have brought about improvement in methods of both breeding and feeding. The analyses of data gathered from these stations have demonstrated the close association of breeding with economic production. By maintaining two constants of environment and nutrition, the influence of genetic factors can be more easily determined.

Pig testing, as at present understood, originated in Denmark, where, in 1896, the first tests of thriftiness and quality were carried out on the farms of certain breeders. In 1907 the first testing station was opened in that country and since then many more have been established.

The technique of testing at all these stations is very similar. A given number of males and females, usually two of each, is drawn from each litter which is to be tested. Great stress is laid on the methods of selection of this „litter group", for it should be as representative as possible of the whole litter. On arrival at the station, the pigs are weighed. They are then maintained at the testing station under identical environmental and nutritional conditions and are slaughtered on reaching a certain weight; during the test the food consumption and live weight gain are noted. The carcases are weighed and the proportions of lean, fat, bone, offal and various cuts are calculated. American methods are an exception in that only two of the four pigs are slaughtered: the remaining two are returned to the farm and used for breeding purposes. There is much to be said in favour of this plan.

From these data, it is possible to obtain the individual or litter live weight gain per day; food consumption per pound of live weight gain; individual litter or breed variations in growth and efficiency, besides much information concerning prolificacy, mothering ability, slaughter loss, quantity and quality of meat produced, and the economics of bacon and pork production.

3. *Pig Recording*

Litter testing as a means for the improvement of the pig may be described as analogous to the laying trial method for the improvement of poultry, since in both cases selected samples are sent to a centre to be tested, under carefully controlled conditions, against similar samples from other breeders. Pig recording, on the other hand, may be compared to milk recording for dairy cattle. The recorder visits the farm and either makes or checks the records, and no attempt is made to standardize the conditions of the different producers. Thus the figures obtained in pig recording are an index of the combined effects of heredity, nutrition and husbandry generally.

In detail, the method of pig recording varies, but it is usual to record the numbers born, numbers and weight at weaning, and the numbers, age and weight when despatched for slaughter, or at some given weight or age. Not infrequently the weight at some pre-weaning age (usually three weeks) is also recorded. Slaughter tests and carcase measurements, as in litter testing, are also frequently made.

HAMMOND (1934a) sums up the results as follows:

„(1) The mothering qualities of the sows and the efficiency of the breeding, feeding and management up to weaning time, which can be measured by taking the weights of the litters at 8 weeks old (and if desired by pedigree breeders at 3 weeks old as well).

(2) The rate of growth to bacon weight, i.e. the efficiency of the system of feeding and management, which can be measured by the number of days taken to reach 200 lbs. live weight (or in days $\pm$ a standard growth curve).

(3) The quality of the carcase produced, which can be measured by graders at the factory who would, if the carcase was below first grade, state the reason why (too fat, too thin in belly, etc.), so that steps could be immediately taken to put it right by changes either in the method of feeding or in the breeding stock."

In that litter testing is the better index of the hereditary qualities of the pig, less emphasis is laid, in the following pages, on pig recording. This, however, does not imply that the authors consider pig recording as of no value. Improvement is the function both of heredity and environment, but this paper is concerned with the genetical aspect.

4. *Current Methods of Litter Testing and Pig Recording*

(*a*) D e n m a r k

Although most breeders are familiar with the actual work carried out at the testing stations, there are still many people who do not fully appreciate the part which they play in the Danish pig improvement scheme. Some time previous to the introduction of testing, the Government had instituted a number of subsidised „Breeding Centres" and, owing to close supervision of these Centres by State officials, very full information relating to the breeding capacity of the stock was obtained by means of detailed private herd registers. In addition to this breeding information, the Litter Testing Stations provided details of carcase quality and the economy of food consumption. Owing to the small size of the country and the manner in which the breeding centre scheme has been organized, practically all the commercial pigs of Denmark are the immediate descendants of animals which have been tested. Thus, the livestock authorities have been able to maintain considerable control over the fecundity,quality and thriftiness of the commercial pigs used throughout the country.

This work has produced information of considerable scientific value, as witness the papers of JESPERSON, MADSEN, LUND, ROSTING and BECK, whose various scientific results have been quoted in their appropriate places in this monograph.

(*b*) S w e d e n

Sweden, largely owing to her greater size and varied geographical conditions, has but recently concentrated upon bacon production to the same extent as Denmark, but is nevertheless fully aware of the value of pig testing and recording. The first attempt to apply such methods resulted in the establishment of a Litter Testing Station at Åstorp in Scania, the southernmost province of the country. This station was modelled very closely upon the Danish ones, and although only completed in 1923, it has already produced several most interesting reports. In general, the methods are the same as those employed in Denmark and much of the information is similarly tabulated. Owing to the fact that the predominant breed in Denmark is the native Landrace, whereas in Sweden over eighty per cent. of the pure-bred pigs are Large Whites descended from stock imported from Great Britain, the Swedish results are not only interesting in themselves but can be applied to a considerable extent to British problems. Among those who are predominantly concerned with the Swedish reports may be mentioned the names of AXELSSON, JOHANSSON, HANSSON, BENGTSSON and LARSSON.

But the position in Sweden is of double interest for those studying the question of type improvement since, in addition to Testing Stations, there has been organised a Pig Recording Scheme. By having both these methods employed simultaneously, the value of each is greatly increased. The criticism of the Litter Testing Stations is that their results can only apply to a small percentage of the population. On the other hand, critics of Pig Recording aver that the interaction of numerous different factors affects the figures obtained, so that it is not easy to distinguish the principal causes responsible for the production of particular records, or to compare pigs raised on one farm with those produced at another. If the information obtained by recording is checked at a testing station, then it can be of the greatest value.

(c) G e r m a n y

For the present no national scheme exists in Germany, but pig recording has been commenced in the two provinces of East Prussia and Hannover. Generally speaking, the methods employed under both schemes are similar, although each province lays more stress on some aspects than on others. In East Prussia recording has been carried out since 1924 by sixteen recording societies with approximately 250 herds and a total of 2,750 recorded sows. The authorities in charge of this scheme have experimented with many different types of records, but have decided that the most important thing from a practical point of view is to obtain as large a number of absolutely reliable records as possible of the one or two important aspects of fecundity.

Results have been described by SCHMIDT, VOGEL, ZIMMERMANN, WILD, SEEDORF and DAHLANDER.

(d) P o l a n d

Poland has now established pig testing stations on the lines of those in Denmark (Rózycki, 1933). Four pigs from each litter are fed on a standard ration up to about 200 lbs. live weight, when they are slaughtered and classified according to bacon quality. The results obtained so far show a lesser degree of uniformity than those found in Scandinavia. It is stated that the Polish pigs are less early maturing than those of other countries.

(e) Great Britain

In England the „Cambridge" Scheme of Pig Recording was in operation during 1928—1931. During the same period in Scotland an experimental testing station was established at Edinburgh. Both of these, having served their purpose, are now defunct. Various counties in England now have recording schemes which it is designed to co-ordinate into a national scheme. In Scotland, the plans for a Testing Station have been proposed, but, owing to the depression, the work is not being pressed forward. Reports on the British work have been written by DAVIDSON, DUCKHAM, KITCHIN, PRICE, SMITH and CALDER.

(f) Canada

For the improvement of Canadian pigs the Government has adopted an advanced registry policy for pure-bred swine. This was inaugurated in 1928 and is not dissimilar to the systems of pig recording which have been established elsewhere. The policy is outlined by ROTHWELL, MACMILLAN and PETERSON (1931), and by ROTHWELL and PETERSON (1934).

The avowed object is „to provide the swine industry with a system of pig testing organized on a national basis". Individual sows are tested for prolificacy, feeding qualities and carcase quality of their progeny. All sows and all their progeny are identified by tattooing. Breeders are required to keep private herd records. Inspections of herds are made by officers of the Dominion Department of Agriculture when pigs are from four to eight weeks of age. The pigs are then tattooed, weighed, and the breeder is required to select five pigs from each litter to be fed for a slaughter test. Four of these, after being reared and finished for market, are shipped to a designated packing plant.

A standard for qualification of the sows is based on three main factors:

(1) Production capacity of the sows;

(2) Capacity of the sows' progeny for early maturity;

(3) Quality of the progeny as revealed in the carcase test.

A sow, to qualify, must obtain a specific standard in each of these respects. Boars may qualify by siring at least three litters, the dams of which have qualified as a result of scores secured through such litters. Details are given concerning the carcase tests and the method of grading. These are distinctly interesting.

Advanced registers have already been published by the Livestock Branch of the Dominion Department of Agriculture for pure-bred swine. The list of sows and boars which qualified during the year ending March 31st, 1933, comprises some 100 names. Full particulars are given concerning the progeny of each sow, particularly as regards the slaughter tests, the points for which are given in detail. It is thus possible for a breeder whose carcases do not come up to standard in one respect to find out from the Register a breeder whose pigs can supply the deficiencies of his own stock.

(g) U. S. A.

A national co-operative performance record scheme is now in opera ion in the United States, organised by the U. S. Department of Agriculture, the National Swine Record of Performance Committee and the agricultural experiment stations, particularly those of Iowa, Minnesota, Wisconsin, West Virginia and Ohio. Litters to be tested must consist at weaning time (56 days) of at least seven pigs if from gilts and of at least eight pigs, if from older sows, and must be from pure-bred parents (i.e. they must be pure-bred or the first cross of two pure-bred animals of different breeds). Four pigs from each litter, two barrows and two gilts, are sold by the breeder to the experiment station at the top Chicago market price. When delivered, they are immunised against hog cholera and subjected to vermifugal treatment. On attaining 72 days of age the test is commenced and the animals are fed rations which are standardised for all lots tested at all the stations participating in the scheme. When the pigs reach 225 lbs. live weight, the two barrows and one gilt are slaughtered, and the value of the carcases is carefully assessed. Thus a record of the rate and economy of live weight gain, feed requirements and the quantity and quality of pork produced is obtained. The breeder is given the opportunity of buying back the remaining gilt at the top Chicago market price.

The National Swine Growers' Association, in conjunction with the National Association of Swine Records, has recently launched a scheme of pig recording. Litters are ear-marked at seven days, are weighed when from 50—62 days of age and again at from 170—190 days. Notes on feeding and management are made. A supervisor checks the results for a maximum of 50 herds for each district.

Those whose names should be mentioned in this connection are, from Iowa, LUSH, CULBERTSON, KILDEE, EVVARD, HELSER and HAMMOND; from Minnesota, FERRIN, ANDERSON and JOHNSON; from Illinois, BULL, OLSON, BIGGER, LONGWELL, LAIBLE and CARROLL; and from Ohio, ROBISON.

5. *Boar Testing*

Another method of improvement is that of boar testing. Work along these lines has been done in Russia. KUDRJAVCEV (1932, 1933) made attempts to determine the genotype of the stud boar. For this purpose he used a method of diallelic and polyallelic crossing, in which two or more males can be tested in two seasons. The method of testing is briefly as follows: A given number of sow groups are chosen to equal the number of boars to be tested. Each sow group is then divided into two sub-groups. In the first season, each boar is mated to the sows of the first sub-group of each group. In the second season, matings are reciprocal, that is each boar is mated to the sows of the second sub-group of the same group. Conditions of feeding and environment are held as constant as possible and a control group of sows is introduced further to eliminate error. This is exactly the method which has been employed for the past four years at the experimental farm of the Institute of Animal Genetics, at the University of Edinburgh.

The writers hold that it is by the employment of this method on as large a scale as possible that a substantial advance will be made in the analysis of the inheritance of the productive qualities of the pig. It also furnishes the only sound method for the progeny testing of boars.

6. *Conclusion*

In the foregoing sections dealing with production and methods of improvement, much has been included which must seem to the purist to be far removed from the science of genetics. Until the productive qualities of the pig have been further analysed, the ultimate synthesis is not possible. There is, however, a demand for the immediate improvement of the hereditary qualities of the pig. The writers claim that the preceding pages show that already there exists knowledge which can profitably be employed to this end. Inevitably, at the present, improvement must take place to a great extent by the method of trial and error. But from the foregoing pages we are persuaded that the time is not far distant when, the analysis of the genetic aspect of the productive qualities of the pig being further advanced, a more scientific synthesis will be possible.

We have considered carefully whether these sections ought properly to be included in a paper which reviews the genetics of the pig. Although these pages do not contain much that is of precise genetic fact, we contend that the papers cited do contain the foundation for the proper genetic analysis of the productive qualities of the pig. Hence, there is no need to apologise for their inclusion.

It must be remembered that it is only within the past five years that attention has been properly directed to this subject. Some seven years ago one of the present authors declined to compile this paper on the ground that there was little information available that was either scientifically accurate or economically valuable. Now the difficulty is to keep pace with all the work which is being done.

Surely these facts are an omen of the part to be played in the future by Science as a means for the improvement of the productive qualities of the pig.

XII. FIGURES 1—21

U.S.A.
Tamworth.
Barrow.

FIG. 1. (Photo. by courtesy of Professor
A. L. Anderson, Iowa State College)

U.S.A.
Poland-China.
Barrow.

FIG. 2. (Photo. by courtesy of Professor
A. L. Anderson, Iowa State College)

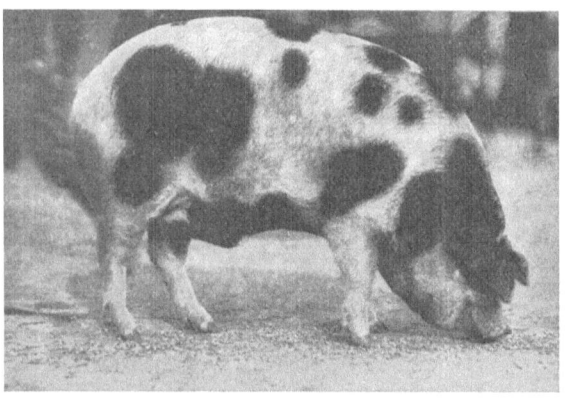

U.S.A.
Spotted
Poland-China.
Barrow.

FIG. 3. (Photo. by courtesy of Professor
A. L. Anderson, Iowa State College)

U.S.A.
Berkshire.
Gilt.

FIG. 4. (Photo. by courtesy of Professor
A. L. Anderson, Iowa State College)

U.S.A.
Hampshire.
Barrow.

FIG. 5. (Photo. by courtesy of Professor
A. L. Anderson, Iowa State College)

U.S.A.
Hampshire boar
which developed
spots at over two
years of age.
See p.

FIG. 6. (Photo. by courtesy of Professor
A. L. Anderson, Iowa State College)

U.S.A.
Hereford.
Piglings.

FIG. 7. (Photo. by courtesy of Professor
J. L. Lush, Iowa State College)

Denmark.
Danish Landrace.
Gilt.

FIG. 8. (Photo. M. Knudsen, by courtesy of
Dr. K. Madsen, Copenhagen)

Germany.
Hannover
Improved
Landschwein.
Hog.

FIG 9. (Photo. by courtesy of Professor H. Nachtsheim,
Inst. f. Vererbungsforschung, Berlin-Dahlem)

Great Britain.
Gloucestershire
Old Spots.
Gilt.

FIG. 10. (Photo. by courtesy of Mr. A. E. Perkins,
Gloucestershire Old Spot Pig Society)

U.S.A.
Duroc-Jersey.
Gilt.

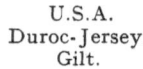

FIG. 11. (Photo. by courtesy of Professor
A. L. Anderson, Iowa State College)

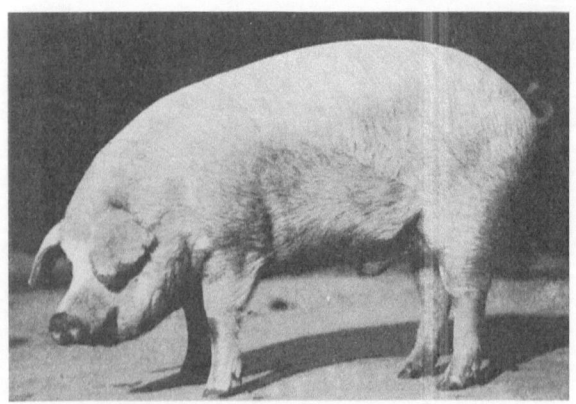

U.S.A.
Chester-White.
Barrow.

FIG. 12. (Photo. by courtesy of Professor
A. L. Anderson, Iowa State College)

Great Britain.
Large Black.
Gilt (6 months).

FIG. 13. (Photo. by courtesy of Mr. B. J. Roche,
Large Black Pig Society)

Great Britain.
Wessex
Saddleback.
Gilt.

FIG. 14. (Photo. by courtesy of Mr. A. Hobson,
National Pig Breeders' Association)

Great Britain.
Berkshire.
Boar.

FIG. 15. (Photo. by courtesy of Mr. A. Hobson,
National Pig Breeders' Association)

Great Britain.
Large White.
Gilt.

FIG. 16. (Photo. by courtesy of Mr. A. Hobson,
National Pig Breeders' Association)

Great Britain.
Middle White.
Gilt (6 months).

FIG. 17. (Farmer & Stock-breeder Photo).

Great Britain.
Tamworth.
Boar.

FIG. 18. (Farmer and Stock-breeder Photo).

Germany.
Güstin Pasture.
Sow.

FIG. 19. (Photo. K. Wittstock, by courtesy of Professor
H. Nachtsheim, Inst. f. Vererbungsforschung,
Berlin-Dahlem).

Germany.
Half-red Bavarian
Landschwein.
Sow.

FIG. 20. (Photo. after Dr. C. Kronacher,
Allgemeine Tierzucht)

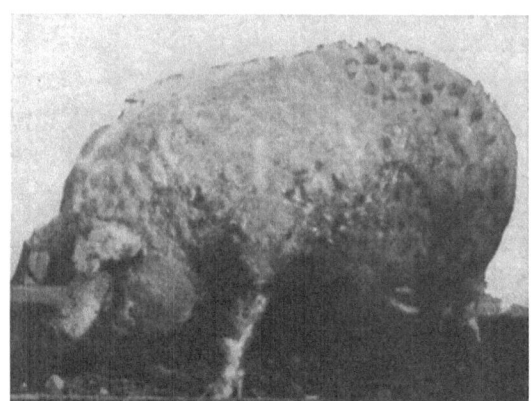

Roumania.
Mangalita.
Sow.

FIG. 21. (Photo. by courtesy of Professor
G. K. Constantinescu, National
Zootechnic Institute, Bucharest)

BIBLIOGRAPHY *).

ABDERHALDEN,E. 1899. Die Beziehungen der Wachstumsgeschwindigkeit des Säuglings zur Zusammensetzung der Milch. Z. phys. Chemie, **27**.

ANONYMOUS. 1923. The value of a first cross in the production of pork and bacon. J. Min. Agric., **29**: 939-941.

ANONYMOUS. 1929. Pig breeding in Sweden. Farmer & Stockbreeder, **43**: 1431—1432.

ANONYMOUS. 1929. Los tipos de cerdos. Riv. Asoc. Argent. Criad. Cerdos, **8**, No. 78: 19—27.

ANONYMOUS. 1931. Litter testing and pig recording. Interim report by the Pig Industry Council. Min. Agric. & Fish., Marketing Leaflet No. 30, 7 pp.

ANONYMOUS. 1931. Leistungskontrolle der Schlesischen Schweinezucht. Ergebnisse der Zuchtleistungsprüfungen Kalenderjahr 1931. Mit einer Durchschnittsberechnung der Kontrolljahre 1926—31. Durchgef. v. d. Landwirtschaftskammer Niederschlesien und dem Verband schlesischer Schweinezüchter E. V.

ANONYMOUS. 1931—32. Ergebnisse der amtlichen Schweine-Leistungs-kontrolle der Landwirtschaftskammer Brandenburg 1931—32.

ANONYMOUS. 1932. Schweinezucht und Schweinehaltung und ihre Aussich-ten. Dtsch. landw. Tierz., **36**, No. 22.

ANONYMOUS. 1932. Pig breeding and disease problems. Vet. Rec. N. S., **12**: 1351—1352.

ANTHONY, D. J. 1929. Some variations in the number of ribs in pigs. Vet. J., **85**: 229.

ARISTOTLE. *Historia Animalium* II. 1—7.

ARMSTRONG, J. T. 1932a. Breeding pigs for export. Tasmanian J. Agric.,**3**: 25—29.

ARMSTRONG, J. T. 193b. The pig industry. Tasmanian J. Agric., **3**: 63—67.

ASDELL, S. A. and SMITH, A. D. Buchanan. 1927. Some practical obser-vations on the inheritance of colour, beard, and tassels. Brit. Goat Soc. Year Book, 4 pp.

ASHLEY, R. C. and MALCOMSON, A. W. 1920. Variation of individual pigs in economy of gains. J. Agric. Res,. **19**: 225—234.

AULD, R. C. 1889. Some cases of solid-hoofed hogs and two-toed horses. Amer. Nat., **23**: 447—449.

AXELSSON, J. 1928. Några betydelstulla faktorer inom vår svinavel. Nor-disk Jordbrugsforsk. : 217—252.

* t.t. = translated title.

AXELSSON, J. 1929a. Resultaten vid renavel och korsning inom svinaveln i Malmöhus län. Svenska Svinavelsför. tidsk., No. 6: 168—181.

AXELSSON, J. 1929b. Sambandet mellan antalet grisar i kullarna vid födseln och medelvikten pr gris i desamma vid olika ålder. Nordisk Jordbrugsforsk.: 123—141.

AXELSSON, J. 1929c. Resulteten från den av Malmöhus Läns svinavelsförening anordnade smågriskontrollen vid Svalöf åren 1927—1928. Förlagsaktiebolagets i Malmö Boktryckeri, Malmö.

AXELSSON, J. 1930. Resultat från smågriskontrollen vid Ruhlsdorf i jämförelse med svenska. Svenska Svinavelsför. tidsk. No. 6: 175—184.

AXELSSON, J. 1933. Einige Resultate der Schweinemastkontrolle in Malmöhus Län. Z. Züchtg., B. 28: 157—191.

BAIRD, W. W. 1923. Swine feeding experiments at the Nappan experimental farm. Canada: Dominion Exp. Farms, Nappan Farm Rpt. Supt.: 12—14.

BAKER, J. R. 1924. Sexual abnormalities in pigs. Pig Breeders' Annual, 4: 112—114.

BAKER, J. R. 1925a. On the descended testes of sex-intergrade hogs. Quart. J. Micr. Sci., 69: 689—701.

BAKER, J. R. 1925b. On sex-intergrade pigs: their anatomy, genetics, and developmental physiology. Brit. J. Exp. Biol., 2: 247—263.

BAKER, J. R. 1926a. Asymmetry in hermaphrodite pigs. J. Anat., 60: 374—381.

BAKER, J. R. 1926b. Sex in Man and Animals. Routledge. London, 175 pp.

BAKER, J. R. 1928. A new type of mammalian intersexuality. Brit. J. Exp. Biol., 6: 56—64.

BARTRAM, H. A. 1926. Über die Wurfgrösse beim veredelten Landschwein. Züchtungskunde, 1: 256—269.

BÄSSMANN, W. 1933. Die Schweinezucht in der Mark Brandenburg. Dtsch. landw. Tierz. 37: 341—343.

BATESON, W. 1894. Materials for the study of variation treated with especial regard to discontinuity in the origin of species. Macmillan & Co., London.

BECK, N. 1931. 19de Beretning om sammenlignende Forsøg med Svin fra statsanerkendte Avlscentre. 139te Beretning fra Forsøgslaboratoriet. København. 184 pp.

BECK, N. 1933. 21de Beretning om sammenlignende Forsøg med Svin fra statsanerkendte Avlscentre 150de Beretning fra Forsøgslaboratoriet. København. 195 pp.

BECKER, E. 1896—97. Uber Zwitterbildung beim Schwein. Verh. Phys.-Med. Ges. Würzburg, 30.

BENGTSSON. S. 1923—24. Grissuggans mjölkavkastning och dennas sammansättning. Nordisk Jordbrugsforsk., 5: 355—364.

BERRY, R. A. and O'BRIEN, D. G. 1921. Errors in feeding experiments with cross-bred pigs. J. Agric. Sci., 11: 275—286.

BESTE, K. 1927. Das deutsche veredelte Landschwein Lüneburger Zucht mit besonderer Berücksichtigung des Blutaufbaues. Dissertation, Halle-Wittenberg.

BIRCH, R. R. 1923. Natural and artificial immunity of young pigs to hog cholera. Cornell Vet., **13**: 159—169.

BITTING, A. W. 1897. The fecundity of swine. Indiana Agric. Exp. Sta., Veter. Dept., 10th Ann. Rpt.: 42—46.

BLANC, L. 1893. Étude sur la polydactylie chez les mammifères. Ann. Soc. Linn. Lyon, **40**: 53—88.

BOMHARD, (Dr.) 1927. Die Erbanlagen der unterfränkischen Schweine- zucht. Z. Schweinez., **34**: 773—778.

BONNIER, G. and HALLQVIST, C. 1930. Letalanlag hos Svin. Medd. Inst. f. Husdjursföräd., No. 3: 46—53.

BRAMBELL, F. W. R. 1927. The histology of an hermaphrodite pig and its developmental significance. J. Anat., **63**: 397—407.

BRYDEN, Wm. 1933. The Chromosomes of the Pig. Cytologia, **5**: 149—153.

BUCKINGHAM, J. L. 1930. An hermaphrodite pig. Vet. Rec., **10**: 967.

BULATOVICI, G. T. 1930. Contribuţiuni la studiul cauzelor lipsei de proli- ficitate la rasa Mangaliţa. Dissertation, Bukarest.

BULL, S. and CARROLL, W. E. 1929. Type in swine as related to rate and economy of gain and quality of pork. Illinois Agric. Exp. Sta., Circ. 345, 14 pp.

BULL, S. and LONGWELL, J. H. 1929. Swine Type Studies: II. Type in swine as related to quality of pork. Illinois Agric. Exp. Sta., Bull. 322: 395—490.

BULL, S. and OLSON, F. C. 1931. [Barrows and gilts about equal in pork produced]. Illinois Agric. Exp. Sta., Ann. Rpt.

CALDER, A. 1928. Report on the bacon pig competition at Highland and Agricultural Society's Show, July, 1928. Scottish Agric. Publish. Co., Glasgow, 19 pp.

CALDER, A. and SMITH, A. D. Buchanan. 1928. Pig testing: the results of preliminary work on bacon type. Scot. J. Agric., **11**: 318—325.

CALDER, A. 1930—31. Experiences of pig testing in Scotland. Pig Breeders' Annual, **10**: 74—85.

CALDER, A. 1931. Individuality and performance in the selecting of breeding pigs. Lancs. County Milk Record. Soc. Year Book, 1931.

CALDER, A. 1933—34. Pig breeding in Scotland. Pig Breeders' Annual, **13**: 123—129.

CALIFORNIA AGRICULTURAL EXPERIMENT STATION. 1929—30. Breeding experiments with Berkshire hogs. California Agric. Exp. Sta. Rpt.: 57.

CALINESCU, J. 1929. Rendementul animalelor de macelarie. Rev. stiintel. vet., No. 9: 247.

CAMPBELL, D. M. 1914. Springtime Surgery. 3rd Ed. Chicago. 163 pp.

CANADA: DOMINION DEPARTMENT OF AGRICULTURE, 1923. Federal assist- ance to hog breeding. The Boar Premium Policy for Swine Clubs. Domin. Depart. Agric., Live Stock Branch. 8 pp.

CARLYLE, see RICHTER, HEMPEL, OHLIGMACHER and RODEWALD, 1928.

CARMICHAEL, W. J. and RICE, J. B. 1920. Variations in farrow: with special reference to the birth weight of pigs. Illinois Agric. Exp. Sta., Bull. 226: 65—95.

CARR-SAUNDERS, A. M. 1922. Note on inheritance in swine. Science, **55**: 19.

CARROLL, W. E. and ROBERTS, E. 1928. Crossbred pigs prove no better than pure-breds. Illinois Agric. Exp. Sta. Rpt.: 124.

CARROLL, W. E., BULL, S., RICE, J. B., LAIBLE, R. J. and SMITH, R. A. 1929. Swine Type Studies: I. Type in swine as related to rate and economy of gain. Illinois Agric. Exp. Sta., Bull. 321: 339—392.

CHLEBAROFF, G. S. 1925—26. [Growth, fattening capacity and economic value of Bulgarian breeds of pig.] Yearb. Agric. Faculty, Sofia Univ.

CHRISTENSEN, F. W., THOMPSON, O. A. and JORGENSON, L. 1926a. Bacon and lard type hogs compared. Production of bacon hog. North Dakota Agric. Exp. Sta., Bull. 194: 23—27.

CHRISTENSEN, F. W., THOMPSON, O. A. and JORGENSON, L. 1926b. Prolificacy of sows and mortality of pigs. North Dakota Agric. Exp. Sta., Bull. 194: 83.

CLAUS, G. 1928. Züchterische Betrachtungen zur veredelten Landschweinezucht. Dtsch. landw. Tierz. **32**, No. 47.

CLEMENTE, L. S. 1931. A genetic study of inbreeding and outbreeding in Berkshire swine. Natur. appl. Sci. Bull., **1**: 173—186.

COLE, L. J., PARK, J. S. and DEAKIN, A. 1933. ,,Seedy Cut'' as affecting bacon production. Wisconsin Agric. Exp. Sta., Res. Bull. 118, 61 pp.

CONGRÈS DE L'ÉLÉVAGE ET L'ALIMENTATION DU PORC. 1928. Travaux Soc. nat. d'encourag. à l'agric., Paris, 559 pp.

CONSTANTINESCU, G. K 1928. Ein rezessives Weiss beim Schwein. Z. indukt. Abst. Vererbg., **47**: 147—150.

CONSTANTINESCU, G. K. 1933. Vererbungsversuche an Schweinen unter besonderer Berücksichtigung des Mangalitzaschweines. Z. Züchtg., B. **26**: 395—427.

COOPER, T. P. 1914. Pork production contest. Breeders' Gaz., **65**: 523—524.

CORNER, G. W. 1920. A case of true lateral hermaphroditism in a pig with functional ovary. Carneg. Instit. Publ. 274: 137—142.

CORNER, G. W. 1921. Internal migration of the ovum. Johns Hopkins Hosp. Bull. 32: 78—83.

COUES, E. 1878. On a breed of solid-hoofed pigs apparently established in Texas. Bull. U. S. Geol. & Geog. Surv., **4**: 295—298.

CRAFT, W.A. 1924—26. Genetic investigations with swine at the Oklahoma Station: A study of two different systems of breeding, limited inbreeding and outcrossing, when practised with swine. Oklahoma Agric. Exp. Sta., Bien. Rpt.: 34—37.

CRAFT, W. A. 1927—30. Outbred hogs outgain inbred hogs. Oklahoma Agric. Exp. Sta. Rpt.: 78—81, 83—85.

CRAFT, W. A. 1927—30. Pigs inherit swirl hair. Oklahoma Agric. Exp. Sta. Rpt.: 86.

CRAFT, W. A. 1929. The influence of birth weight upon subsequent development of inbred and outbred pigs. Proc. Amer. Soc. Anim.Prod.: 128—130. (1930).

CRAFT, W. A. 1931. Further observations on inbred and outbred pigs in the feed yard. Proc. Amer. Soc. Anim. Prod.: 265—266. (1932).

CREW, F. A. E. 1923a. Three cases of developmental intersexuality in the pig. Veter. J., **79**, 6 pp.

CREW, F. A. E. 1923b. Studies in intersexuality. I. A peculiar type of developmental intersexuality in the male of the domesticated mammals. Proc. Roy. Soc., B. **95**: 90—109.

CREW, F. A. E. 1924a. Blue and white colour in swine. J. Hered., **15**: 395—396.

CREW, F. A. E. 1924b. Hermaphroditism in the pig. J. Obst. Gynaecol., **31**: 369—386.

CREW, F. A. E. 1924c. Genetics and the pig breeder. Pig Breeders' Annual, **4**: 40 —45.

CREW, F. A. E. 1925. Prenatal death in the pig and its effect upon the sex-ratio. Proc. Roy. Soc. Edin., **46**: 9—14.

CREW, F. A. E. 1927. Abnormal sexuality in animals. II. Physiological. Quart. Rev. Biol., **2**: 249—266.

CREW, F. A. E. 1927. The genetics of sexuality in animals. Cambridge University Press.

CULBERTSON, C. C. and EVVARD, J. M. 1925. The costly influence of an inferior sire. Iowa Agric. Exp. Sta., Leaflet 1, 7 pp.

CULBERTSON, C. C. and EVVARD, J. M. 1925—26. A superior litter of crossbred pigs. Proc. Amer. Soc. Anim. Prod.: 53—59. (1927).

CULBERTSON, C. C., EVVARD, J. M., KILDEE, H. H. and HELSER, M. D. 1928. Swine Performance Record — Litter Comparisons. Series I. Iowa Agric. Exp. Sta., Leaflet 26, 14 pp.

CULBERTSON, C. C., EVVARD, J. M., KILDEE, H. H. and HELSER, M. D. 1931. Swine Performance Record — Litter Comparisons. Iowa Agric. Exp. Sta., Bull. 277: 86—115.

CULBERTSON, C. C., KILDEE, H. H., HELSER, M. D. and HAMMOND, W. E. 1932. Swine Performance Record — Litter Comparisons. — Series II. Iowa Agric. Exp. Sta., Leaflet 28, 8 pp.

CULBERTSON, C. C., KILDEE, H. H., HELSER, M. D. and HAMMOND, W. E. 1933. Swine Performance Record — Litter Comparisons — Series III. Iowa Agric. Exp. Sta., Leaflet 30, 8 pp.

DABROWA-SZREMOWICZ, S. V. 1905a. Eine neue Abart von Schweinen. Illust. Landw. Ztg., **25**: 564—565.

DABROWA-SZREMOWICZ, S. V. 1905b. Einhuferschweine. Illust. Landw. Ztg., **25**: 810—811.

DAHLANDER, — 1927. Erste Leistungsprüfung der ostpreussischen Schweinezüchter-Vereinigung Königsberg i. Pr. Dtsch. landw. Tierz., **31**, No. 45.

DAHLANDER, G. 1932. Die Zucht des Schweines. Anleit. dtsch. Ges. Zchtngsk., Heft 19.

DAMLE, D. M. 1931. How many times wild pigs breed in one year? Agric. & Livestock in India, 1: 274—275.

DANISH FOREIGN OFFICE JOURNAL, 1930. Pig breeding and the bacon industry. The Danish Foreign Office, Copenhagen, November, 1930.

DARWIN, C. 1868. The variation of animals and plants under domestication. John Murray, London.

DASSOGNO, L. 1915. Sulla durata della gravidanza nelle scrofe Yorkshires. Industr. Latt. Zootec., No. 12: 180—182.

DAVID, L. T. 1932. Histology of the skin of the Mexican hairless swine (*Sus scrofa*). Amer. J. Anat., **50**: 283—292.

DAVIDSON, H. R. 1924. Records of production and their use in breeding. Pig Breeders' Annual, **4**: 26—30.

DAVIDSON, H. R. 1927. Variations in carcase type for pork and bacon. Scot. J. Agric., **10**: 394—403.

DAVIDSON, H. R. 1930. Reproductive disturbances caused by feeding protein-deficient and calcium-deficient rations to breeding pigs. J. Agric Sci., **20**: 233—264.

DAVIDSON, H. R. 1933. The results of twenty-five years of pig testing in Denmark and their relation to pig recording methods in Great Britain. Pig Breeders' Annual, **13**: 34—45.

DAVIDSON, H. R. 1933. Pig recording as a factor in cheapening and increasing pig production. J. Min. Agric., **40**: 42—50.

DAVIDSON, H. R. 1934. First cross pigs for bacon. Farmer & Stockbreeder, N. S., **48**: 2552.

DAVIDSON, H. R. and SMITH, A. D. Buchanan. 1927—28. Pig testing and recording for the purposes of advanced registry. Pig Breeders' Annual, **7**: 3—19.

DAVIDSON, H. R. and SMITH, A. D. Buchanan. 1927—28. Recording and pig testing for the purpose of advanced registry. Pig Breeders' Annual, **7**: 46—62.

DAVIDSON, H. R. and DUCKHAM, M. A. 1929. First report on the East Anglian pig recording scheme. Dept. Agric., University of Cambridge.

DAY, G. E. 1915. Productive swine husbandry. Lippincott's Farm Manuals, 2nd Edition; J. B. Lippincott Company, Philadelphia & London. 363 pp.

DEAKIN, A. 1932a. Melanic pigmentation of the mammary glands of the black breeds and the red breed of pigs. Proc. VI. Intern. Cong. Genetics, **2**: 41—42.

DEAKIN, A. 1932b. Cases of extreme sexual infantilism in the sow. Anat. Rec., **51**: 361—371.

DECHAMBRE, P. 1925. L'Hybridation du sanglier et du porc. Rev. d'hist. nat. appl., **6**: 207—212.

DECHAMBRE, P. 1926. La reproduction entre les porcs et les sangliers, hybridation ou croisement? Rev. Zootech., **5**, No. 4.

DECHAMBRE, P. 1929. Études génétiques sur les porcs et les sangliers. Réc. méd. vétér., **105**: 129—184.

DECHAMBRE, P. 1934. La croissance du jeune porc pendant son élevage. Rev. Zootech., No. 2: 75—79.

DEIGHTON, T. 1929. A study of the metabolism of two breeds of pigs. J. Agric. Sci., **19**: 140—184.

DETLEFSON, J. A. & CARMICHAEL, W. J. 1921. Inheritance of syndactylism, black and dilution in swine. J. Agric. Res., **20**: 595—604.

DIECKERT, F. 1929. Über die Fruchtbarkeit des weissen deutschen Edelschweines nach Untersuchungen an Herdbuchtieren der ostpreussischen Schweinezüchtervereinigung Allenstein in den Jahren 1918—27. Züchtungskunde, **4**: 474—483.

DÖHRMANN, H. 1930. Alkalmas-e az isohaemogglutinatio és a precipitatio a sertés fajtatisztaságak megallapitására? Dissertation, Budapest.

DRAHN, F. 1923. Zur Entstehung der Hyperdaktylie beim Schwein. Eine embryologisch-entwicklungs-mechanische Studie. Arch. wiss. prakt. Tierhlk., **49**: 245—260.

DUCHANEK, J. O. 1894. Hermaphroditismus bei Schweinen. Tierärztl. Zbl., **17**.

DUCKHAM, A. N. 1928—29. The commercial value of pig recording. Pig Breeders' Annual, **8**: 67—72.

DUCKHAM, A. N. 1929a. Second report on the East Anglian pig recording scheme. Cambridge.

DUCKHAM, A. N. 1929b. Pig recording in East Anglia. J. Min. Agric., **36**: 731—738.

DUCKHAM, A. N. 1929c. Aspects of pig recording. Essex County Farmers' Yearbook.

DUCKHAM, A, N. 1929—30. The interpretation of pig recording results. Pig Breeders' Annual, **9**: 70—76.

DUCKHAM, A. N. 1930—31. Rationalising the pig industry. Pig Breeders' Annual, **10**.

DUNCKER, G. 1915. Die Frequenzverteilung der Geschlechtskombinationen bei Mehrlingsgeburten des Menschen und des Schweins. Biol. Zbl., **35**: 506—539.

DUNLOP, G. 1933. Methods of experimentation in animal nutrition. J. Agric. Sci., **23**: 580—614.

DURHAM, G. B. 1921. Inheritance of belting spotting in cattle and swine. Amer. Nat., **55**: 476—477.

ELLINGER, T. 1921. The influence of age on fertility in swine. Proc. Nat. Acad. Sci., **7**: 134—138.

ELLINGER, T. & EVVARD, J. M. 1923. The register of performance for swine. Proc. Amer. Soc. Anim. Prod.: 57—61. (1924).

EUDES-DESLONGCHAMPS. 1842. Mém. Soc. Linn. de Normandie, **7**: 41. (Cited by Darwin).

EVANS, B. R. 1930. Duroc Sentinel, **5**, No. 11: 3.

EVVARD, J. M. & CULBERTSON, C. C. 1926. Studies of swine types. Amer. Herdsman, No. 1: 6.

EVVARD, J. M. 1926. Crossbred hogs win new laurels. Amer. Swineherd, No. 12: 4.

EVVARD, J. M., CULBERTSON, C. C., SHEARER, P. S., BASSETT, C. F. & HAM-
MOND, W. E. 1926. Part II. Development of the crossbred and pure-
bred pigs from weaning time to market. Iowa Agric. Exp. Sta., Leaflet No. 20.

EYTON, T. C. 1837. Proc. Zool. Soc. Lond., **5**: 23.

FARGO, J. M. and COYNER, J. M. 1930. Experiments with swine in Alabama. Pigs that make best gains also yield best carcases. Alabama Agric.
Exp. Sta. Rpt.: 24, 25.

FEIGE, E. 1932. Die Nutzung unserer Schweinerassen. Berl. tierärztl.
Wschr., **48**: 716—717..

FERRIN, E. F. and McCARTY, M. A. 1923. The comparative feed requirements and rate and cost of gains of fall and spring farrowed pigs. Proc.
Amer. Soc. Anim, Prod.: 71—74 (1924).

FERRIN, E. F. 1930. Testing breeding hogs for big production. Swine World,
17 (4).

FERRIN, E. F. 1931. Selecting the most profitable brood sow type. Swine
World, **18**.

FERRIN, E. F., ANDERSON, P. A. and JOHNSON, Do. 1931. The selection of
breeding stock on the basis of economy of gains and carcass values.
Minnesota Agric. Exp. Sta., Private Publication.

FERRIN, E. F., ANDERSON, P. A. and JOHNSON, Do. 1932. National swine
record of performance. Minnesota Agric. Exp. Sta., Private Publication.

FERRIN, E. F. 1932. Production tests for the selection of breeding hogs.
Proc. Amer. Soc. Anim. Prod.: 134—137 (1933).

FEUNTEUN, L. 1932. Étude zootechnique de la production porcine en Co-
chinchine. Bull. Écon. Indo-Chine. (Rev. Zootech. No. 12: 101—409).

FINLAY, G. F. 1924. Sterility in swine. Nat. Council Pig Breed. & Feed.,
Bull. No. 1, 24 pp.

FISHBEIN, M. 1913. Isoagglutination in man and lower animals. J. Infect.
Diseases, **12**: 133—139.

FLEMING, G. 1902. A textbook of operative veterinary surgery. Vol. 2. London & New York.

FÖRSTER, —. 1927. Die Eberstationen als Mittel zur Hebung der Schweinezucht. Dtsch. landw. Presse, **54**: 661.

FRINGS, P. 1932. Die Zucht des veredelten Landschweines in der Provinz
Schleswig-Holstein. Arb. Dtsch. Ges. Zchtngskde., Heft 54, 77 pp.

FRÖLICH, G. 1911. Fruchtbarkeit und Geschlechtsverhältnis beim weissen
Edelschwein. Jahrb. wiss. prakt. Tierzucht, **6**: 451—454.

FRÖLICH, G. 1913. Uber die Ergebnisse experimenteller Vererbungsstudien
beim Schwein. J. Landw., **61**: 217—235.

FRÖLICH, G. 1919. Über Abstammung und Inzucht, an der Hand der wichtigsten Blutlinien des weissen deutschen Edelschweins, Ammerländer.
Dtsch. landw. Tierz., **23**: 69—71, 76—78, 83—85.

FRÖLICH, G. 1927. Zucht- und Nutztypen in der Schweinezucht. Z. f.
Schweinez., **34**: 73—77.

136   THE GENETICS OF THE PIG

FROST, J. A. 1933. (In litteris) 11.2.33.

FUNKQUIST, H. 1927. Der Stand der Rinder-, Schaf- und Schweinezucht in Schweden und ihre Hebung. Züchtungskunde, 2: 289—312.

FUNKQUIST, H. 1929. Erbliche Begattungsunfähigkeit bei Zuchtebern. Hereditas, 13: 107—120.

GADEAU DE KERVILLE, H. 1902. Notes de Tératologie mammalogique et ornithologique. Bull. Soc. Amis Sci. Nat. Rouen: 131—143.

GALBUSERA, S. 1920. Di alcuni ibridi osservati in Sardegna. La clinica veter., 43: 385—386.

GARTH, W. 1894. Zwei Fälle von Hermaphroditismus verus beim Schwein. Dissertation, Giessen, 59 pp.

GÄRTNER, R. 1931. Vererbung beim Schwein. Züchtungskunde, 6: 241—249.

GENTRY, N. W. 1905. Inbreeding Berkshires. Amer. Breed. Ass. Rpt.: 168—171.

GEORGE, C. R. 1912. The inheritance of fecundity in Poland-China sows. Thesis, University of Ohio.

GINIEIS, —. Hermaphrodisme chez le porc. Rec. Méd. vétér. Alfort, 85: 478—481.

GIRARD, A. 1921. Les races porcines méridionales. Rev. vétér. (Toulouse), 73: 82—95, 466—485.

GODBEY, E. G. 1930. A study of the results of intensively inbreeding Berkshire swine. South Carolina Agric. Exp. Sta. Rpt.: 35.

GOELDI, E. A. 1914. Ueber atavistische Längsstreifung bei den neugeborenen Jungen gewisser Rassen des Hausschweines. C. R. 9. Congr. Internat. Zool.: 369—370.

GOHREN. See RICHTER, HEMPEL, OHLIGMACHER and RODEWALD, 1928.

GOLF, A. 1927. Neuere Forschungsergebnisse in der Schweinezucht und Schweineaufzucht. Mitt. dtsch. Landw.-Ges., 42, No. 49: 1108—1112.

GONZALEZ, B. M. & LAGO, F. P. 1923. Improving Philippine swine. J. Philippine Agric., 12.

GOULD, H. N. 1923. Observations on the genital organs of a sex-intergrade hog. Anat. Rec., 26: 241—261.

GRAHAM, R., TUNNICLIFF, E. A. & FRANK, E. R. 1925. [Livestock disease investigations]. Illinois Sta. Rpt.

GRAHAM, R. & TUNNICLIFF, E. A. 1926. [Investigations on swine diseases at the Illinois Station]. Illinois Sta. Rpt.

GRAHAM, R., TUNNICLIFF, E. A. & McCULLOCH, E. C. 1927. [Work with hog cholera at the Illinois Station]. Illinois Sta. Rpt.: 88—90.

GRAHL, W. 1904. Acht Fälle von Zwitterbildung beim Schwein, darunter ein Fall von Hermaphroditismus verus lateralis. Inaug. Dissert., München. pp. 1—49.

GRANDI, W. C. 1931. Pig-keeping costs. Min. Agric. & Fish., Bull. 33: 1—26.

GRIMES, J. C. & SEWELL, W. E. 1928. Improving scrub hogs by the use of purebred sires. Alabama Agric. Exp. Sta. Rpt.: 12—13.

GRIMES, J. C., SEWELL, W. E. & TAYLOR, W. C. 1929. Improving scrub hogs by the use of purebred sires. Alabama Agric. Exp. Sta. Rpt.: 13—15.

GRIMES, J. C. & SEWELL, W. E. 1930. Experiments with swine in Alabama. Alabama Agric. Exp. Sta., Rpt.

GRIMES, J. C., SEWELL, W, E. & TAYLOR, W. C. 1930. Grading up hogs by the use of purebred sires. Alabama Agric. Exp. Sta., Bull. 234, 12 pp.

GRISWOLD, D. J., TROWBRIDGE, P. F., HOGAN, A. G., & HAIGH, L. D. 1928. The effect of gestation and lactation upon the growth and composition of swine. Missouri Agric. Exp. Sta., Res. Bull. 114, 62 pp.

GUTBROD, K. 1932. Die Rasse- und Typfrage in der süddeutschen Schweinezucht. Südd. Landw. Tierzucht, 26, No. 34.

HABU, Y. 1930. Inheritance of face shape in swine. Japan. J. Genet., 6: 143—167.

HABU, Y. 1932. Comparison of the growth of swine with special reference to the utilisation of F_1. Imp. Zootech. Exp. Sta. Chiba-Shi., Bull. No. 30, 36 pp.

HAECKER, V. 1918. Entwicklungsgeschichtliche Eigenschaftsanalyse (Phaenogenetik). Gemeinsame Aufgaben der Entwicklungsgeschichte, Vererbungs- und Rassenlehre. Fischer, Jena, 344 pp., 181 figs.

HALDANE, J. B. S. 1927. The comparative genetics of colour in rodents and carnivora. Biol. Rev., 2: 199—212.

HALE, F. 1933. Pigs born without eye balls. J. Hered., 24: 105—106.

HALLAM, 1833. [Letter on a singular race of pigs.] Proc. Zool. Soc. London, p. 16.

HALLQVIST, C. 1933. Ein Fall von Letalfaktoren beim Schwein. Hereditas, 18: 215—224.

HAMMOND, J. A. 1912. Case of hermaphroditism in the pig. J. Anat. Physiol., 7: 307—312.

HAMMOND, J. 1914. On some factors controlling fertility in domestic animals. J. Agric. Sci., 6: 263—277.

HAMMOND, J. 1921. Further observations on the factors controlling fertility and foetal atrophy. J. Agric. Sci., 11: 337—366.

HAMMOND, J. 1922. On the relative growth and development of various breeds and crosses of pigs. J. Agric. Sci., 12: 387—423.

HAMMOND, J. 1926. The importance of a prolific strain. Pig Breeders' Annual, 6: 73—75.

HAMMOND, J. 1927. Growth and conformation in the pig. Pig Breeders' Annual, 7: 76—81.

HAMMOND, J. 1929. Fertility in pigs. Pig Breeders' Annual, 9: 23—27.

HAMMOND, J. 1932. Pigs for pork and pigs for bacon. J. Roy. Agric. Soc. England, 93: 131—145.

HAMMOND, J. 1933. The anatomy of pigs in relation to market requirements. Pig Breeders' Annual, 13: 28—33.

HAMMOND, J. 1934a. Conformation and development of the bacon pig. Agric. Progress, 11: 127—129.

HANCE, R. T. 1917—18. The diploid chromosome complexes of the pig (Sus scrofa) and their variations. J. Morph., 30: 155—222.

HANCE, R. T. 1918. Variations in somatic chromosomes. Biol. Bull., 35: 33—37.

Hansson, N. & Bengtsson, S. 1923. Verksamheten vid försöksstationen för genomförande av avkastningskontroll inom svinaveln under år 1923. Medd. Centralanst. försöksväs. jordbr., No. 266, Husdjuravd. No. 40, 24 pp.

Hansson, N. & Bengtsson, S. 1924. Verksamheten vid försöksstationen för avkastningskontroll inom svinaveln under år 1924. Medd. Centralanst försöksväs. jordbr., No. 289, Husdjuravd. No. 46, 78 pp.

Hansson, N. & Bengtsson, S. 1926. Verksamheten vid försöksstationen för avkastningskontroll inom svinaveln under år 1925. Medd. Centralanst. försöksväs. jordbr., No. 306, Husdjuravd. No. 49.

Hansson, N. 1927a. Leistungskontrolle in der schwedischen Schweinezucht. Z. Tierz. Zücht Biol., 10 : 341—376.

Hansson, N. 1927b. De svenska svinerasernas fruktsamhet. Landtmannen: 589—590.

Hansson, N. & Bengtsson, S. 1927. Verksamheten vid försöksstationen för avkastningskontroll inom svinaveln under år 1926. Jämte översikt över de under åren 1923—1926 uppnådda medelresultaten. Medd. Centralanst. försöksväs. jordbr., No. 322, Husdjuravd. No. 53, 79 pp.

Hansson, N. & Bengtsson, S. 1928. Verksamheten vid försöksstationen för avkastningskontroll inom svinaveln under år 1927. Medd. Centralanst. försöksväs. jordbr., No. 339, Husdjuravd. No. 59, 53 pp.

Hansson, N. & Bengtsson, S. 1929a. Verksamheten vid försöksstationen för avkastningskontroll inom svinaveln under år 1928. Medd. Centralanst. försöksväs. jordbr., No. 363, Husdjuravd. No. 64, 31 pp.

Hansson, N. & Bengtsson, S. 1929b. Översikt över resultaten av Åstorpsförsöken åren 1923—28. Medd. Centralanst. försöksväs. jordbr., No. 364, Husdjuravd. No. 65, 50 pp.

Hansson, N. & Bengtsson, S. 1930. Verksamheten vid försöksstationen för avkastningskontroll inom svinaveln under år 1929. Medd. Centralanst. försöksväs. jordbr., No. 381, Husdjuravd. No. 73, 43 pp.

Hansson, N. & Bengtsson, S. 1931. Verksamheten vid försöksstationen för avkastningskontroll inom svinaveln under år 1930. Medd. Centralanst. försöksväs. jordbr., No. 402, Husdjuravd. No. 77, 64 pp.

Hansson, N. & Bengtsson, S. 1932. Verksamheten vid försöksstationerna för avkastningskontroll inom svinaveln under år 1931. Medd. Centralanst. försöksväs jordbr., No. 414, Husdjuravd. No. 80, 55 pp.

Harris, J. A. 1916. Variation, correlation, and inheritance of fertility in the mammals. Amer. Nat., 50 : 626—636.

Hart, E. B. & Steenbock, H. 1918a. Hairless pigs. Wisconsin Agric. Exp. Sta., Bull. 297, 11 pp.

Hart, E. B. & Steenbock, H. 1918b. Thyroid hyperplasia and the relations of iodine to the hairless pig malady. J. Biol. Chem., 33 : 313—322.

Hays, F. A. 1919. Inbreeding animals. Delaware Agric. Exp. Sta., Bull. 123: 5—49.

Hayward, H. 1917. Inbreeding experiments with pigs. Delaware Agric. Exp. Sta. Rpt.: 17.

HEMPEL, K. 1925. Über die Entwicklung der Ferkel während der Säugezeit und die Einrichtung von Leistungsprüfungen in der Schweinezucht.
Dissertation, Göttingen.

HENNING, G. F. & STOUT, W. B. 1932. Factors influencing the dressing
percentage of hogs. Ohio Agric. Exp. Sta., Bull. 505, 32 pp.

HERLYN, K. E. 1928. Über Blutgruppen bei Tieren. Züchtungskunde, 3:
377—398.

HERMANN, G. 1932. Über die Schilddrüsen von Wild- und Hausschwein
und ihren verschiedenen Rhythmus in der Tätigkeit. Arch. Tierheilk.,
64: 547—550.

HICKS, W. H. 1922. Experiments with swine at the Agassiz experimental
farm. Canada Exp. Farms, Agassiz Farm Rpt. Supt.: 15—18.

HICKS, W. H. et alii. 1924. Experiments with swine at the Canadian Experimental Farms. Canada Exp. Farms, Rpts. Supts.

HICKS, J. S. 1933. Genetics and the practical breeder of pure-bred stock.
Pig Breeders' Annual, 13: 45—50.

HINMAN, R. B. 1926. Bacon breeds. Amer. Swineherd, 3: 16,

HOBDAY, F. T. G. 1914. Castration and ovariotomy. 2nd Ed., 162 pp. W. &
A. K. Johnston Ltd., Edinburgh & London.

HOBSON, A. 1931. Wessex Saddleback markings. Nat. Pig Breed. Ass.
Gaz., No. 16: 37—46.

HÖFLIGER. 1931. Haarkleid und Haut des Wildschweines, VII. Beitrag zur
Anatomie von Sus scrofa L. und zum Domestikationsproblem. Z. Anat.,
96: 551—623.

HOFMAN-BANG, N. O. 1922. 10de Beretning om sammenlignende Forsøg med
Svin fra statsunderstøttede Avlscentre. 109de Beretning fra Forsøgslaboratoriet. København. 109 pp.

HOGAN, A. G. 1923. The relation of age and weight of swine to gains in
nutrients. Proc. Amer. Soc. Anim. Prod.: 28—31. (1924).

HOGAN, A. G. & McKENZIE, F. F. 1926. Fecundity of swine: The normal
sexual cycle and the cycle as influenced by age and by unfavourable
dietary conditions. Missouri Agric. Exp. Sta., Bull. 244: 24.

L'HOSTIS, J. & GILBERT, M. 1923. Un cas d'hermaphrodisme chez le porc.
Rec. Méd. Vétér. Alfort, 100: 427—435.

HUGHES, E. H. 1933. Inbreeding Berkshire swine. J. Hered., 24:
199-203.

HUGHES, E. H. 1935. Polydactyly in swine. J. Hered., 26: 415—418.

HUGHES, E. H. & HART, H. 1934. Defective skulls inherited in swine.
J. Hered., 25: 111—115.

HUGHES, W. 1927. Sex-intergrades in foetal pigs. Biol. Bull., 52: 121-137.

HUGHES, W. 1929. The free-martin condition in swine. Anat. Rec., 41:
213—245.

HUGHES, W. 1930. The twinning condition in swine. Trans. R. Canad. Inst.,
17: 209—216.

HUTT, F. B. 1934. Inherited lethal characters in domestic animals. Cornell
Veterinarian, 24: 1—25.

IDAHO AGRICULTURAL EXPERIMENT STATION. 1931. Genetic studies with swine at the Idaho Station. Idaho Agric. Exp. Sta., Bull. 179: 20.

ILLINOIS AGRICULTURAL EXPERIMENT STATION. 1923. Illinois Agric. Exp. Sta. Rpt.: 16.

ILLINOIS AGRICULTURAL EXPERIMENT STATION. 1923—24. [Seek strain of hogs resistant to cholera.] Illinois Agric. Exp. Sta. Rpt.: 72.

ILLINOIS AGRICULTURAL EXPERIMENT STATION. 1924—25. [Intermediate pigs still superior in type tests.] Illinois Agric. Exp. Sta. Rpt.: 55—57.

ILLINOIS AGRICULTURAL EXPERIMENT STATION. 1926. Illinois Agric. Exp. Sta. Rpt.: 49—59.

INDIANA AGRICULTURAL EXPERIMENT STATION. 1930. Swine efficiency test completed. Indiana Agric. Exp. Sta. Rpt.

IOWA AGRICULTURAL EXPERIMENT STATION. 1918. [Different types of swine respond differently to feed.] Iowa Agric. Exp. Sta. Rpt.: 20—21.

IOWA AGRICULTURAL EXPERIMENT STATION. 1922. [The question of swine type.] Iowa Agric. Exp. Sta. Rpt.: 23.

IOWA AGRICULTURAL EXPERIMENT STATION. 1925. [Type test with swine.] Iowa Agric. Exp. Sta. Rpt.: 62—63.

IOWA AGRICULTURAL EXPERIMENT STATION. 1926. [Swine inbreeding versus outbreeding.] Iowa Agric. Exp. Sta. Rpt.: 23.

IOWA AGRICULTURAL EXPERIMENT STATION. 1926. [Breeding swine for natural resistance to cholera.] Iowa Agric. Exp. Sta. Rpt.: 43.

IOWA AGRICULTURAL EXPERIMENT STATION. 1927. [Swine feeding and breeding experiments at the Iowa Station.] Iowa Agric. Exp. Sta. Rpt.: 17, 18, and 21.

IOWA AGRICULTURAL EXPERIMENT STATION. 1928a. [Investigations with swine at the Iowa Station. Type tests with swine.] Iowa Agric. Exp. Sta. Rpt.: 48, 49.

IOWA AGRICULTURAL EXPERIMENT STATION. 1928b. [Breeding for resistance to cholera in swine.] Iowa Agric. Exp. Sta. Rpt.: 33—34.

IOWA AGRICULTURAL EXPERIMENT STATION. 1929. [Breeding for resistanc to hog cholera at the Iowa Station.] Iowa Agric. Exp. Sta. Rpt.: 33—34.

IOWA AGRICULTURAL EXPERIMENT STATION. 1930. [Swine performance record.] Iowa Agric. Exp. Sta. Rpt.: 52—53.

IOWA AGRICULTURAL EXPERIMENT STATION. 1931. [Experiments with livestock.] Iowa Agric. Exp. Sta. Rpt.: 22—31, 33—35, 37.

IOWA AGRICULTURAL EXPERIMENT STATION. 1933. Pig testing results. Iowa Agric. Exp. Sta., Leaflet 30.

IVANOV, M. F. 1933. Novaja poroda svineĭ - Ukrainskaja belaja. Probl. Životn., No. 1: 32—42.

JAKOBIEC, J. and MARCHLEWSKI,T. 1932. O wpływie bliskiego chowu krewniaczego na biologiczne i użytkowe właścuvości świni domowej. Bull. Acad. Polon. Sci. Lett., B. II: 347—361.

JESPERSEN, J. and MADSEN, M. P. O. 1929. Beretning om Afkomsundersøgelser over Orner af Dansk Landrace. Bull. Roy. Vet. Agric. Coll., Copenhagen.

JESPERSEN, J. & MADSEN, M. P. O. 1931. Beretning om Afkomsundersøgelser over Orner af Dansk Landrace₁ Samvirk. Danske Andelsslag., 138 pp.

JOEST, E. 1921. Spezielle pathologische Anatomie der Haustiere. 2, 650 pp. Berlin.

JOHANSSON, I. 1929a. Statistiska undersokningar över svinens fruktsamhet. Nordisk Jordbrugsforsk.: 85—99.

JOHANSSON, I. 1929b. Statistische Untersuchungen über die Fruchtbarkeit der Schweine. Z. Tierzücht. Zücht Biol., 15: 49—86.

JOHANSSON, I. 1929c. Dräktighetsidens längd hos nötkreatur och svin. Ultuna Lantbruksinst. årsberät.: 41—59.

JOHANSSON, I. 1931. Problems in breeding for high prolificacy. Pig Breeders' Annual, 11: 80—87.

JORGENSON, C. E. 1922. Field and laboratory studies of infectious enteritis in young pigs and the efficiency of bacterins in the control of this disease. N. Amer. Vet., 3: 642—646.

JOSEPH, W. E. 1924. Feeding brood sows and growing the litters. Montana Agric. Exp. Sta., Bull. 165, 18 pp.

KAEMPFFER, A. 1932. Über ein zweites Isoagglutinogen-Agglutininpaar, B-B, im Schweineblut. Z. Rassenphysiol., 5: 53—58.

KAEMPFFER, A. 1932. Über die Vererbung der Blutgruppen des Schweins. Z. indukt. Abst. Vererbg., 61: 261—293.

KALUGIN, I. I. 1925. Contributions to the study of tri- and polydactylous pigs of White Russia, II, III, IV, and V. (t.t.) Bull. White Russ. State Inst. Agric. & Forest., No. 5, 30 pp., No. 7, 65 pp., No. 8, 53 pp. + 47 plates.

KANSAS AGRICULTURAL EXPERIMENT STATION. 1915—16. Inheritance investigation in swine. Kansas Agric. Exp. Sta. Rpt.: 19.

KANSAS AGRICULTURAL EXPERIMENT STATION. 1917—18. Inheritance investigations in swine. Kansas Agric. Exp. Sta. Rpt.: 42—43.

KAYSER, W. 1929. Individualitätsreaktionen des Blutes von Schafen, Ziegen, Schweinen und Rindern. Dissertation, Berlin.

KEITH, T. B. 1930. Relation of size of swine litters to age of dam and to size of succeeding litters. J. Agric. Res., 41: 593—600.

KENNESSEY, B. 1930. Adatok a Magyarországi Berkshiretényesztéshez. Mezögazdasági Közlöny, Budapest, 5 sez.: 237—246.

KENTUCKY AGRICULTURAL EXPERIMENT STATION. 1922. Size, sex and weight of pigs per litter. Kentucky Agric. Exp. Sta. Rpt., pt. 1: 46—47.

KING, F. G. 1926. Is motherliness in brood sows an inheritable character? Poland China J., 12, No. 17: 4.

KINGSBURY, B. F. 1909. Report of a case of hermaphroditism (H. verus lateralis) in Sus scrofa. Anat. Rec., 3: 278—282.

KINZELBACH, W. 1931. Untersuchungen über Atresia ani beim Schweine, Z. indukt. Abst. Vererbg., 60: 84—124.

KITCHIN, A. W. M. 1930. Third report on the East Anglian pig recording scheme. Animal Nutrition Institute, University of Cambridge.

KITCHIN, A. W. M. 1931. Economic aspects of pig recording. J. Min. Agric., 37: 957—965.

KLOSS, C. B. 1921. Malaysian bearded pigs. J. Straits Branch, Roy. Asiatic Soc., No. 83: 147—150.

KOCH, P. & NEUMÜLLER, O. 1932. Gesichtsspalten als Erbfehler beim Schwein. Dtsch. tierärztl. Wschr., **40**: 353—357.

KONOPIŃSKI, T. 1932. Poszukiwania współzależności pomiędzy liczbą sutek u macior a ilością urodzonych prosiąt. Rocz. Nauk Roln. Lesn., **28**: 277—300.

KOPSCH, FR. & SZYMONOWICZ, L. 1896. Ein Fall von *Hermaphroditismus verus bilateralis* beim Schweine, nebst Bemerkungen über die Entstehung der Geschlechtsdrüsen aus dem Keimepithel. Anat. Anz., **12**: 129—139.

KOSSWIG, C. & OSSENT, H. P. 1931. Die Vererbung der Haarfarben beim Schwein. Z. Züchtg., B. **22**: 297—383.

KOSSWIG, C. & OSSENT, H. P. 1932. Ein Beitrag zur Vererbung der Haarfarben beim Schwein. Züchter, **4**: 225—230.

KOSSWIG, C. & OSSENT, H. P. 1933. Bemerkungen von Kosswig-Ossent zur Arbeit Kronacher-Ogrizek. Z. Züchtg., B. **26**: 429—430.

KOZELKA, A, W. 1929. The inheritance of natural immunity among animals. J. Hered., **20**: 519—530.

KRALLINGER, H. 1930. Über einige das Geschlechtsverhältnis beeinflussende Faktoren. Züchtungskunde, **5**: 490—502.

KRALLINGER, H. 1931. Cytologische Studien an einigen Haussäugetieren. Arch. Tierern. u. Tierz., **5**: 127—187.

KRALLINGER, H. & SCHOTT, A. 1933. Untersuchungen über Geschlechtsleben und Fortpflanzung der Haustiere. II. Der Einfluss des Decktermins innerhalb der Rausche der Schweine auf ihre Fruchtbarkeit und das Geschletsverhältnis der Nachkommen. Arch. Tierern u. Tierz., **9**: 41—49.

KRALLINGER, H. & SCHOTT, A. 1934. Untersuchungen über Geschlechtsleben und Fortpflanzung der Haustiere. III. Hat der Eber einen Einfluss auf die durchschnittliche Ferkelzahl der von ihm erzeugten Würfe? Züchtungskunde, **9**: 175—179.

KREDIET, G. 1929. Gonade und Uterus beim intersexuellen Schwein. Baum-Festschr. Schaper, Hannover.

KREDIET, G. 1930. Hermaphroditismus. Geneesk., **28**: 199—282.

KREDIET, G. 1932. Ein Genitalapparat eines intersexuellen Schweines mit 2 Ovariotestes. Tijdschr. Diergeneesk., **59**: 194—202.

KREINER, H. 1926. Das halbrote bayrische Landschwein im Entwicklungsgang der Schweinezucht in Altbayern, besonders in Niederbayern und in der Oberpfalz. Land. i. Bayern, Rietsch, Gm. b. H.

KRONACHER, C. 1924. Vererbungsversuche und Beobachtungen an Schweinen. Z. indukt. Abst. Vererbg., **34**: 1—120.

KRONACHER, C. 1926. Nochmals: Schweineleistungsprüfungen. Mitt. dtsch. Landw.-Ges., **41**: 943—948.

KRONACHER, C. 1927. Allgemeine Tierzucht. Part IV. Die Züchtung. Paul Parey, Berlin: 549—550.

KRONACHER, C. 1930. Weitere Vererbungsbeobachtungen und -Versuche an Schweinen. Z. Züchtg., B. **18**: 315—365.

KRONACHER, C. & OGRIZEK, A. 1932. Vererbungsversuche und -Beobachtungen an Schweinen. III. Z. Züchtg., B, **25**: 3—43.

KRONACHER, C. & OGRIZEK, A. 1933. Bemerkungen Kronacher-Ogrizek. Stellungnahme zu den Bemerkungen C. Kosswigs zu unserer Arbeit „Vererbungsversuche und -Beobachtungen an Schweinen. III". Z. Züchtg., B, **26**: 431—434.

KRÜGER, H. 1926. Schlachtbeobachtungen und Ausschlachtungsversuche an Schweinen. Wiss. Arch. Landw., **3**.

KUCHENBECKER, L. 1931. Ein Schwein mit Eselhufen. Dtsch. landw. Presse, **58**: 310.

KUDRJAVCEV, P. N. 1932. Opyt ocenki summarnogo genotipa proizvoditeleĭ v svinovodstve. Probl. Životn., No. 7: 82—87.

KUDRJAVCEV, P. N. 1933. Polialleljnoe skreščivanie kak metod ispytanija hrjakov. Probl. Životn., No. 4.: 42—46.

KUHLMAN, A. H. & COLE, L. J. 1929. Birth weight important factor in swine production. Wisconsin Agric. Exp. Sta., Bull. 405: 64.

KULOW, H. 1928. Ein Vergleich zweier Blutlinien von Wurfgeschwistern des deutschen veredelten Landschweines in bezug auf Körperformen und Leistungen. Z. Tierzüchtg. ZüchtBiol., **13**: 1—92.

LACY, M. D. 1932. Differences between barrows and gilts in the proportion of pork cuts. Proc. Amer. Soc. Anim. Prod.: 354—357. (1933).

LAIBLE, R. J., BULL, S. & MITCHELL, H. H. 1924. Swine type tests favour intermediate pig. Illinois Agric. Exp. Sta. Rpt.: 59—69.

LAIBLE, R. J., SMITH, R. A. BULL, S. & LONGWELL, J. H. 1925. Intermediate pigs still superior in type tests. Illinois Agric. Exp. Sta. Rpt.: 55—57.

LAMBERT, W. V., MURRAY, C. & SHEARER, P. S. 1928. Selection for natural resistance to cholera in swine. Proc. Amer. Soc. Anim. Prod.: 33—37. (1929).

LARSSON, S. 1928. Gödningsgradens inverkan på svinens Foderförbrukning pr kg. tillväxt. Medd. Centralanst. försöksväs. jordbr., No. 338, Husdjuravd. No. 58, 26 pp.

LARSSON, S. 1929. Recording and litter testing in Sweden. Pig Breeders' Annual, **9**: 28—33.

LAYLEY, G. W. & MALDEN, W. J. 1935. The evolution of the British pig. London: John Bale, Sons & Danielsson, Ltd. 119 pp.

LEENY, H. 1920. Livestock of the Farm. London.

LEHMANN, F. 1911. Über den Futterbedarf und die Leistung des säugenden Mutterschweines. J. Landw., **59**: 317—363.

LEHMANN, F. 1911. Grundlagen der modernen Schweinemast. Jahrb. dtsch. Landw.-Ges., **16**: 942—955.

LEHMANN, W. 1924. Studien über den Zusammenhang von Wüchsigkeit mit Trockensubstanz und Alkaligehalt des Blutes bei Schweinen. Dissertation, Bern.

LETARD, E. & LEGENDRE, G. 1933. Situation de l'élevage porcin en France. Rev. Zootech., No. 4: 300—310.

LLOYD-JONES, O. & EVVARD, J. M. 1919. Studies on color in swine. Iowa Agric. Exp. Sta., Res. Bull. 53: 203—208.

LOWREY, L. G. 1911. Prenatal growth of the pig. Amer. J. Anat., 12: 107—138.

LUCEY, J. 1931—32. Pig breeding in the Irish Free State. Pig Breeders' Annual, 11: 121—125.

LUND, A., BECK, H. & ROSTING, P. 1925. 13de Beretning om sammenlignende Forsøg med Svin fra statsunderstøttede Avlscentre. 117de Beretning fra Forsøgslaboratoriet. København, 146 pp.

LUSH, J. L. 1921. Inheritance in swine. J. Hered., 12: 57—71.

LUSH, J. L. 1923. Cross breeding of swine and the chief results. Breed. Gaz., 83: 74.

LUSH, J. L. 1925. The possibility of sex control by artificial insemination with centrifuged spermatozoa. J. Agric. Res., 30: 893—913.

LUSH, J. L. 1926. Inheritance of horns, wattles, and color. J. Hered., 17: 72—91.

LUSH, J. L. 1931. Genetic aspects of the record of performance work with swine. Proc. Amer. Soc. Anim. Prod.: 51—62. (1932).

LUSH, J. L. & CULBERTSON, C. C. 1931. The consequences of inbreeding Poland-China hogs. Iowa Agric. Exp. Sta. Rpt.: 23.

LUSH, J. L., CULBERTSON, C. C. & HAMMOND, J. A. 1931. Weight at birth important in pigs. Poland China J., 18, No. 4.

LÜTHGE, H. 1930. Das deutsche veredelte Landschwein. Z. Schweinez., No. 36: 599—601.

LÜTHGE, H. 1933. Beobachtungen über die Bedeutung der Ausgeglichenheit von Ferkelwürfen und die Vererbung dieser Eigenschaft. Züchtungskunde 8: 333—344.

MACHENS, A. 1915. Fruchtbarkeit und Geschlechtsverhältnis beim veredelten Landschwein. Berl. tierärztl. Wchschr., 31: 559—562.

MacCARTHY, C. J. 1929. The department's policy in regard to the improvement of cattle and swine. Dept. Agric. J., No. 2: 135—148. (Dublin).

MACKENZIE, K. J. J. & MARSHALL, F. H. A. 1915. On ovariotomy in sows; with observations on the mammary glands and the internal genital organs. J. Agric. Sci., 7: 243—245.

MALSBURG, K. 1924. Recherches génétiques sur le porc monongulé. Rozp. Biolog.: 247—254.

MANZANO, 1934. Performing pigs. Farmer & Stockbreeder, Jan. 8th.: 94.

MARSHALL, L. M. 1928. Standardising quality in pigs. Live Stock J.: 107.

MASKO, R. 1930. Über das Blutbild beim Hausschwein. Dissertation, Zagreb.

MATTHEWS, B. 1927. Leistungskontrolle in der Schweinezucht. Dtsch. landw. Tierzucht, 31, No. 46.

McARTHUR, C. L. 1918. The immunity of suckling pigs to hog cholera. Arkansas Agric. Exp. Sta. Bull. 151: 3—22.

McKENZIE, F. F. 1928. Growth and reproduction in swine. Missouri Agric. Exp. Sta., Res. Bull. 118, 67 pp.

McKenzie, F. F. 1931. [Studies on physiology of reproduction]; Cryptorchidism in swine. Missouri Agric. Exp. Sta., Bull. 300.

McKenzie, F. F. 1932. [Breeding experiments at the Missouri Experiment Station]. Missouri Agric. Exp. Sta., Bull. 310.

McLean, J. A. 1914. The sapphire hog. J. Hered., 5: 301—304.

McPhee, H. C. & Zeller, J. H. 1925. Unusual coat colours in swine. J. Hered., 16: 347—350.

McPhee, H. C. 1925—26. The value of herd book data for the genetic investigations of sex-ratio and frequency of sex combinations in swine. Proc. Amer. Soc. Anim. Prod.: 203—206. (1927).

McPhee, H. C. 1927. The swine herd-book as a source of data for the investigation of the sex-ratio and frequency of sex-combinations in pig litters. J. Agric. Res., 34: 715—726.

McPhee, H. C. 1930a. Swine inbreeding at the United States Department of Agriculture Laboratories and at the State Experiment Stations. Proc. Amer. Soc. Anim. Prod.: 128—131. (1931).

McPhee, H. C. 1930b. Swine inbreeding at the United States Department of Agriculture — a Progress Report. Proc. Amer. Soc. Anim. Prod.: 131—134. (1931).

McPhee, H. C. 1931. Size of litter as a selection index in swine. Proc. Amer. Soc. Anim. Prod.: 262—264. (1932).

McPhee, H. C., Russell, E. Z. & Zeller, J. 1931. An inbreeding experiment with Poland-China swine. J. Hered., 22: 393—403.

McPhee, H. C. 1932. The effects of inbreeding and crossbreeding on swine. Proc. VI. Intern. Cong. Genetics, 2: 132—133.

McPhee, H. C. & Buckley, S. S. 1934. Inheritance of cryptorchidism in swine. J. Hered., 25: 295—303.

Meiners, F. 1922. Sieben Fälle von Polydactylie beim Schwein. Med. vet. Dissertation, Berlin.

Melvin, A. D. & Schroeder, E. C. 1906. Animal breeding and disease. U. S. Dept. Agric., Bur. Anim. Indust., Rpt.: 213—222.

Miller, P. E. 1919. Hairless pigs. Minnesota Agric. Exp. Sta., Morris Substa. Rpt.: 45—46.

Milne, B. C. 1920. Experiments with swine in Canada. Canada Exp. Farms Rpt.: 172.

Milovanov, V. K. 1932. Sovremennoe sostojanie voprosa iskusstvennogo osemenenija svinei. Probl. Životn., No. 4: 31—34.

Minkler, F. C. 1916. Hog cholera. New Jersey Agric. Exp. Sta. Rpt.: 132—136.

Mohler, J. R. 1933. Report of the Chief of the Bureau of Animal Industry, 1933. U. S. Dept. Agric., Bureau Anim. Ind., 47 pp.

Mohr, O. 1930. Dødbringende arvefaktorer hos husdyr og mennesker. Naturens Verden., 14: 1—31.

Möller, H. & Dollar, J. A. W. 1903. The practice of veterinary surgery. Vol. 3, Regional surgery. 583 pp. London.

Morkeberg, P. A. 1916. [The present position and future prospects of swine breeding in Denmark.] Tiddskr. Landökonomi, **5**: 233—269, **6**: 324—336; Dept. Agric. Tech. Inst. Ireland J., **17**: 40—56.

MORKEBERG, P. A. 1926. Selection methods in Denmark. Pig Breeders' Annual, **6**: 48—56.

MORRIS, H. P. & JOHNSON, D. W. 1932. Effects of nutrition and heredity upon litter size in swine and rats. J. Agric. Res., **44**: 511—521.

MORRISON, R. 1926. The individuality of the pig; its breeding, feeding and management. London, John Murray: xii + 377 pp.

MOSKOVITS, F. 1931. Present tendencies in European pig production. Month. Bull. Agric. Sci. & Prac., **22**: 225—231.

MUMFORD, F. B. 1921. The effect on growth of breeding immature animals. Missouri Agric. Exp. Sta., Res. Bull. 45: 3—37.

MUMFORD, F. B. & BERNARD, P. M. 1924. Age as a factor in animal breeding. Missouri Agric. Exp. Sta., Bull. 210: 34—36.

MUMFORD, F. B. & McKENZIE, F. F. 1925. Age as a factor in animal breeding. Missouri Agric. Exp. Sta., Bull. 228: 31.

MUMFORD, F. B., HOGAN, A. G. & McKENZIE, F. F. 1925—26. Swine breeding; the effect of age, unfavourable dietary conditions and the normal oestrous cycle. Proc. Amer. Soc. Anim. Prod.: 85—88. (1927).

MUND, F. 1926. Die Entwicklung der Zucht des veredelten Landschweins in der Provinz Sachsen. Arb. Dtsch. Ges. Zchtngsk., No. 31.

MURRAY, G. N. 1934. A statistical analysis of growth and carcase measurements of baconers. Onderstepoort J. Vet. Sci., **2**: 301—360.

NACHTSHEIM, H. 1922. Vererbung bei Schweinen. I. Schweine als Versuchstiere für Vererbungsexperimente. Z. Schweinez., **29**: 65—71.

NACHTSHEIM, H. 1924. Über die Vererbung der Zitzenzahl beim Schwein und ihre Bedeuting für die Praxis. Mitt. dtsch. Landw.-Ges., **39**: 455—459.

NACHTSHEIM, H. 1924. Vererbungversuche an Schweinen: Die Vererbung von Farbe und Zeichnung. Z. indukt. Abst. Vererbg., **33**: 317.

NACHTSHEIM, H. 1924. Vererbungsversuche an Schweinen: Die Vererbung der Zitzenzahl. Z. indukt. Abst. Vererbg., **33**: 307—311.

NACHTSHEIM, H. 1925a. Zitzenzahl und Ferkelzahl. Dtsch. landw. Tierzucht, **29**: 345—349.

NACHTSHEIM, H. 1925b. Untersuchungen über Variation und Vererbung des Gesäuges beim Schwein. Z. Tierzücht. ZüchtBiol., **2**: 113—161.

NACHTSHEIM, H. 1933. Zur Frage der Züchtung seuche-immuner Hausschweine. Züchtungskunde, **8**: 450—452.

NATHUSIUS, — 1864. „Schweinschädel". Berlin. (as cited by Darwin, 1868).

NATHUSIUS, [SIMON von.] 1912. Hochinteressante Vererbung bei Schweinen. Ill. landw. Ztg., **33**: 618.

NEUHAUS, M. 1930. Inheritance of Number of Nipples in the Domestic Pig. (t.t.) Ž. eksper. Biol. (Russian), **6**: 73—77.

NEUHAUS, M. 1930. Inheritance of Some Characters in the Pig. (t.t.) Ž. eksper. Biol. (Russian), **6**: 233—236.

NEW ZEALAND DEPARTMENT OF SCIENTIFIC AND INDUSTRIAL RESEARCH,

1930. Pig recording, bacon manufacture, and some aspects of the pig industry in New Zealand. New Zeal. Dep. Sci. Indust. Res., Bull. 17, 62 pp. Wellington.

NORDBY, J. E. 1929a. An inherited skull defect in swine. J. Hered., 20: 229—232.

NORDBY, J. E. 1929b. Congenital skin, ear and skull defects in a pig. Anat. Rec., 42: 267—280.

NORDBY, J. E. 1929c. Sterility in boar responsible for reduced size of litters. J. Amer. Vet. Med. Ass., 74: 911—914.

NORDBY, J. E. 1930. Congenital ear and skull defect in swine. J. Hered., 21: 499—501.

NORDBY, J. E. 1932a. Inheritance of whorls in the hair of swine. J. Hered. 23: 397—404.

NORDBY, J. E. 1932b. Type in market swine and its influence on quality of pork. Idaho Agric. Exp. Sta., Bull. 190, 8 pp.

NORDBY, J. E. 1933a. Congenital melanotic skin tumors in swine. J. Hered., 24: 361—364.

NORDBY, J. E. 1933b. Cryptorchidism and its economic importance to the producer of swine and the processor of pork products. J. Amer. Vet. Med. Ass., 82: 901—912.

NORDBY, J. E. 1934a. White-spotting in Duroc-Jersey swine. J. Agric. Res., 49: 625—634.

NORDBY, J. E. 1934b. Kinky tail in swine. J. Hered., 25: 171—174.

OHIO AGRICULTURAL EXPERIMENT STATION. 1930—31. Fiftieth annual report. Hog experiments. Ohio Agric. Exp. Sta., Bull. 497: 128—132.

OHIO AGRICULTURAL EXPERIMENT STATION. 1931—32. Fifty-first annual report; swine experiments. Ohio Agric. Exp. Sta., Bull. 516: 81 84.

OHLIGMACHER, K. 1925. Untersuchungen über die Entwicklung der Ferkel während der Säugezeit unter besonderer Berücksichtigung der ersten Kreuzungswürfe aus veredelten Landschweinen mit Berkshires. Dissertation, Göttingen.

OKLAHOMA AGRICULTURAL EXPERIMENT STATION. 1924—26. A report on investigations of farm problems. Oklah. Agric. Exp. Sta., Bien Rpt.: 34—37 [See Craft, W. A.]

OSSENT, H. P. 1929. Rezessives Weiss und Frischlingsstreifung der Mangalitza-Schweine. Züchter, 1: 11—13.

OSSENT, H. P. 1932. Ein seuche-immunes wildfarbiges Hausschwein. Züchter, 4: 152—158.

OSSENT, H. P. 1933. Die Züchtung widerstandsfähiger Schweinerassen. Züchtungskunde, 8: 399—402.

OSTERTAG, —. and ZUNTZ, —. 1908. Untersuchungen über die Milchsekretion des Schweines und die Ernährung der Ferkel. Landw. Jahrb., 37: 201—260.

PALMER, C. C. 1917. Partial thyroidectomy in pigs. Amer. J. Physiol., 42: 572—581.

PARK, J. S. 1930. Tests show new factors affecting pork yields. National Provisioner, March 1st, June 7th and August 30th, 1930.

PARKER, G. H. & BULLARD, C. 1900. The arrangement of the mammary glands in litters of unborn pigs. Science, **11**: 168.

PARKER, G. H. & BULLARD, C. 1913. On the size of litters and the number of nipples in swine. Proc. Amer. Acad. Arts Sci., **49**: 399—426.

PARKER, G. H. 1914. A note on sex determination. Science, **39**: 215—216.

PARKES, A. S. 1923a. Studies on the sex-ratio and related phenomena. III. The influence of size of litter. Ann. Appl. Biol., **10**: 287—292.

PARKES, A. S. 1923b. Studies on the sex-ratio and related phenomena. IV. The frequencies of sex-combinations in pig litters. Biometrika, **15**: 373—381.

PARKES, A. S. 1925. Studies on the sex-ratio and related phenomena. VII. The foetal sex-ratio in the pig. J. Agric. Sci., **15**: 284—299.

PARKES, A. S. 1926. Studies on the sex-ratio and related phenomena. VIII. The seasonal sex-ratio in the pig. Z. indukt. Abst. Vererbg., **40**: 121—138.

PATEL, D. S. 1932. How many times wild pigs breed in a year. Agric. & Livestock in India, **2**: 232.

PEARL, R. 1913. On the correlation between the number of mammae of the dam and size of litter in mammals. I. Interracial correlation. II. Intraracial correlation in swine. Proc. Soc. Exp. Biol. Med., **11**: 27—32.

PEARL, R. 1918. The seasonal distribution of swine breeding. Science Month., **7**: 244—251.

PEASE, M. S. 1929. Mendelism. Pig Breeders' Annual, **9**: 44—48.

PERGOLI, — 1924. Le odierne conoscenze sull' eredità e la loro pratica applicazione zootecnica. Giorn. Agricolt. del Domenica, **34**: 60.

PETERSEN, P. & OETKEN, Fr. 1896. Untersuchungen über die Zusammensetzung der Schweinemilch, speciell über den Fettgehalt derselben. Milchztg.: 665, 736.

PICK, L. 1914. Über den wahren Hermaphroditismus des Menschen und der Säugetiere. Arch. mikr. Anat., **84**: 119—242.

PLAIM, R. 1930. Über Messungen am Becken weiblicher Schweine verschiedener Rassen. Vet.-med. Dissertation, Wien.

PLUMB, C. S. 1927. Marketing farm animals. The Athenaeum Press, Ginn and Company, Boston, U. S. A., 366 pp.

POPOV, N. A. 1926. The influence of age on the intensity and cost of fattening pigs (t. t.). Trans. Yaroslav. Zootech. Exp. Sta.: 122—124.

PRENTISS, C. W. 1902—03. Polydactylism in man and the domestic animals with especial reference to digital variation in swine. Bull. Mus. Comp. Zool., **40**: 245—314.

PRICE, W. T. 1932. Pig keeping in Scandinavia. Agric. Prog., **9**: 78—84.

PRICE, W. T. 1932. Pig recording in Wiltshire. Pig Breeders' Anuval, **12**: 102—107.

PÜTZ, H. 1889. Ein Fall von *Hermaphroditismus verus unilateralis* bei einem Schweine. Dtsch. Z. Tiermed., **15**: 91—100.

RACZ, M. 1931. Problems and methods of breeding lard pigs. Month. Bull. Agric. Sci. & Pract., **22**: 467—474.

RANG, F. 1931. Untersuchungen über die Isohämagglutination im Blute des Schweines und Rindes mit eingeengten Seren. Dissertation, Göttingen.

REED, F. H. 1922. Swine experiments at the Lacombe experimental station. Canada Expt. Farms, Lacombe Sta., Rpt. Supt.: 38—48.

REED, F. H. & CHAPMAN, L. T. 1927. Swine husbandry in Central Alberta. Canada Dept. Agric., Bull. 73, N. S., 32 pp.

REYNOLDS, M. H. 1910. Immunity in young pigs from cholera immune sows. Amer. Vet. Rev., **38**: 236—237.

RHOAD, A..O. 1934. Woolly hair in swine. J. Hered., **25**: 371—375.

RICHTER, K. 1925..Die gewichtsmässige Entwicklung der Ferkel der veredelten Landschwein- und der deutschen weissen Edelschweinrasse während einer zehnwöchigen Säugeperiode. Dissertation, Göttingen.

RICHTER, K. 1926. Zwillings- und Mehrlingsgeburten bei unseren landwirtschaftlichen Haussäugetieren. Arb. dtsch. Ges. Zchtngsk., No. 29: 88—100.

RICHTER, K., HEMPEL, K., OHLIGMACHER, K. & RODEWALD, A. 1928. Untersuchungen an Sauen und Ferkeln während der Säugezeit bei den wichtigsten deutschen Schweinerassen. Arb. dtsch. Ges. Zchtngsk., No. 37, 81 pp.

RITZOFFY, N. 1931. Prinos k poznavanju turopoljskog svinjćeta. (English Summary). Vet. Arh.: 83—134.

RITZOFFY, N. 1933. Die Rolle der Inzucht in der Turopoljer Schweinerasse. Z. Züchtg. B., **27**: 419—429.

ROBERTS, E. & RICE, J. B. 1924. Illinois Agric. Exp. Sta. Rpt.: 70—73.

ROBERTS, E. 1925—26. Inheritance of resistance to disease. Proc. Amer. Soc. Anim. Prod.: 50—53. (1927).

ROBERTS, E. & LAIBLE, R. J. 1925. Heterosis in pigs. J. Hered., **16**: 383—385.

ROBERTS, E. 1925. Heredity involved in hairlessness of animals. Illinois Agric. Exp. Sta. Rpt.: 59.

ROBERTS, E. & CARD, L. E. 1925—26. Further data on inheritance of resistance to disease. Proc. Amer. Soc. Anim. Prod.: 201—203. (1927).

ROBERTS, E. 1927. Study of hairlessness now directed at causes. Illinois Agric. Exp. Sta. Rpt.: 74.

ROBERTS, E. CARROLL, W. E. 1927. [Work with hog cholera at the Illinois Station.] Illinois Agric. Exp. Sta. Rpt.: 88—90.

ROBERTS, E. 1928. Heredity looms larger in animal abnormalities. Illinois Agric. Exp. Sta. Rpt.: 124—125.

ROBERTS, E. & CARROLL, W. E. 1931. The inheritance of ,,hairlessness'' in swine-Hypotrichosis. II. J. Hered., **22**: 125—132.

ROBERTS, J. A. FRASER. 1929. The inheritance of a lethal muscle contracture in the sheep. J. Genet., **21**: 57—59.

ROBERTSON, G. S. 1933. Pig breeding and marketing in Northern Ireland. Pig Breeders' Annual, **13** 108—115.

ROBINSON, H. G 1922. Evolution of the Tamworth, Large White and Middle White pigs. Pig Breeders' Annual, **2**: 15—24.

ROBINSON, H. G. 1923. Some digressions upon breeding. Pig Breeders' Annual, **3**: 7—11.

ROBINSON, H. G. 1934. May on the Farm. J. Min. Agric., **41**: 192—194.

ROBISON, W. L. 1919. Effect of age of pigs on rate and economy of gains. Ohio Agric. Exp. Sta., Bull. 335: 545—575.

ROBISON, W. L. 1932. Comparative gains of purebred and crossbred hogs. Ohio Agric. Exp. Sta., Bull. 497: 130.

ROBISON, W. L. 1932. Dressed yields of barrows and sows. Ohio Agric. Exp. Sta., Bull. 497: 129—130.

RODEWALD, J. 1926. Über die Milchleistung der Sauen und das Wachstum der Ferkel des hann.-braunschw. Landschweines während der acht-wöchigen Säugezeit. Dissertation, Göttingen.

ROMMEL, G. M. 1906. The fecundity of Poland-China and Duroc-Jersey sows. U. S. Dept. Agric., Circ. 95, 12 pp.

ROMMEL, G. M. 1907. Inheritance of litter size in Poland-China sows. Amer. Breed. Ass. Rpt., **3**: 201—208.

ROMMEL, G. M. & PHILLIP, E. F. 1907. Inheritance in female line of size of litter in Poland-China sows. Proc. Amer. Phil. Soc., **45**: 245—254.

ROTHWELL, G. B. 1920. Experiments with swine in Canada. Canada Expt. Farms Rpt.: 190.

ROTHWELL, G. B. 1921. Experiments with swine. Canada Expt. Farms, Anim. Husb. Div., Interim Rpt.: 26—39.

ROTHWELL, G. B. 1924. Feeding experiments with swine at the Central Experimental Farm. Canada Expt. Farms, Anim. Husb. Div., Rpt.: 36—52.

ROTHWELL, G. B., MACMILLAN, A. A., and PETERSON, A. W. 1931. The advanced registry policy for pure-bred swine. Live Stock Branch, Dominion Dept. Agric., Ottawa, Canada, 10 pp.

ROTHWELL, G. B. and PETERSON, A. W. 1934. The advanced registry policy for pure-bred swine. Live Stock Branch, Dominion Dept. Agric., Ottawa, Canada, 4 pp.

RÓŻYCKI, K. 1933. Dotychczasowe wyniki badań kontrolnych nad trzodą bekonową. Przegląd hodowl., **7**: 175—180.

RUSSELL, E. Z. 1922. Breeds of swine. U. S. Dept. Agric., Farm Bull. 1263, 22 pp.

SABATINI. 1908. Untersuchungen über die Dauer der Trächtigkeit bei unseren wichtigsten Haustieren beinflusst durch Frühreife, Erstgeburt, sowie Zahl und Geschlecht der Föten. Jahrb. wiss. prakt. Tierzucht, **3**.

SANDERS, H. G. 1931. Sows' milk yields. J. Min. Agric., **37**: 1041—1042.

SAUERBECK, E. 1909. Über den *Hermaphroditismus verus* und den Hermaphroditismus im allgemeinen von morphologischen Standpunkt aus. Frankf. Z. Pathol., **3**: 339—357, 661—705, 829—878.

SCHERMER, S. 1929. Über das Vorkommen von Blutgruppen bei den Haustieren, zugleich ein Beitrag zur Frage der Grollschen Agglutinationsbilder. Züchtungskunde, **4**: 169—179.

SCHERMER, S. 1930. Die Blutgruppen bei den Haustieren. Forsch. Fortschr., **6**: 366—367.

SCHERMER, S. KAYSER, W., & KAEMPFFER, A. 1930. Vergleichende Untersuchungen über die Isoagglutinine im Blute des Menschen und des Schweines. Z. Immunitätsfrschg., **68**: 437—449.

SCHERMER, S. & KAEMPFFER, A. 1932. Über die genetische Bedingtheit der Isoagglutinine auf Grund von Blutgruppenuntersuchungen beim Schwein. Klin. Wchschr., **11**: 335—336.

SCHMIDT, See Richter, Hempel, Ohligmacher and Rodewald, 1928.

SCHMIDT, Ad. 1915. Über den Einfluss der Domestikation auf die mechanischen Qualitäten der *Pars compacta* von *Sus scrofa dom.* nebst einigen Beiträgen zur Theorie der funktionellen Anpassung des Extremitätenskeletts. Arch. Entw. Mech., **41**: 472—534, 605—671.

SCHMIDT, J. & LAUPRECHT, E. 1926. Über die Milch der veredelten Landschweinsauen und ihre Zusammensetzung. Züchtungskunde, **1**: 50—62.

SCHMIDT, J., LAUPRECHT, E. & VOGEL, H. 1926. Beiträge zur Entwicklung und Ernährung der Ferkel während der Säugezeit. Züchtungskunde, **1**: 242—256.

SCHMIDT, J. 1927. Über Leistungsprüfungen in der Schweinezucht, mit besonderer Berücksichtigung der Provinz Hannover. Dtsch. landw. Presse, **54**: 658—659.

SCHMIDT, J. & VOGEL, H. 1927. Bericht über die Schweineleistungsprüfungen der Landwirtschaftskammer für die Provinz Hannover. Z.. Schweinez., **34**: 521—528.

SCHMIDT, J. & VOGEL, H. 1928. Leistungsprüfungen an veredelten Landschweinen auf dem Versuchsgut Friedland der Universität Göttingen. Züchtungskunde, **3**: 153—182.

SCHMIDT, J., VOGEL, H. & ZIMMERMANN, C. 1929. Leistungsprüfungen an deutschen veredelten Landschweinen und deutschen weissen Edelschweinen. Arb. dtsch. Ges. Zchtngsk., No. 47, 146 pp.

SCHMIDT, J. & VOGEL, H. 1930a. Bericht über die Schweineleistungsprüfungen in der Provinz Hannover. Züchtungskunde, **5**: 193—216.

SCHMIDT, J. & VOGEL, H. 1930b. Ein Beitrag zur Rassenbeurteilung des veredelten Landschweines und des deutschen Edelschweines auf Grund ihrer Nutzleistungen. Dtsch. landw. Presse, No. 2: 16.

SCHOENFELD, —. 1928. Leistungsprüfungen bei Schweinen. Dtsch. landw. Tierzucht, **32**, No. 16.

SCHOESSAU, W. 1927. Die Leistungsprüfungen in Schlesien. Z. Schweinez., **34**: 703—707.

SCHOTT, A. 1931. Studien über die züchterische Bedeutung der Blutgruppen des Schweines. Wiss. Arch. Landw., B. **7**: 68—108.

SCHOTTERER, A. 1933. Polydaktylie und Gaumenspalte bei deutschen Edelschweinen. Z. Züchtg., B. **26**: 219—223.

SCHWEDER, G. 1895. [Note on the Occurrence of three-toed pigs on the Upper Dvina.] Korresp.bl. Naturf. Ver. Riga, **38**: 82.

SCHWYZER, M. 1928. Die Fruchtbarkeit des veredelten Landschweins. Schweiz. landw. Monatshefte, **6**: 251—253.

SEEDORF, —. 1932. Nachrichten über den Vieh-und Fleischmarkt. No. 44.

SEVERSON, A. 1917. Color inheritance in swine. J. Hered., **8**: 379—381.

SEVERSON, A. 1925—26. Prolificacy of sows and mortality of pigs. Proc. Amer. Soc. Anim. Prod.: 60—62. (1927).

SEVERSON, A. 1932. [Experiments with livestock in North Dakota.] North Dakota Agric. Exp. Sta., Bull. 256.

SHAW, A. M. 1929. Variations in the skeletal structure of the pig. Sci. Agric., **10**: 23—27.

SHAW, A. M. 1930. A method of determining the variations in the vertebral column of the live pig. Sci. Agric., **10**: 690—695.

SHEARER, P. S., EVVARD, J. M., & CULBERTSON, C. C. 1926. Crossbreds *versus* purebreds in producing market hogs. Iowa Agric. Exp. Sta., Leaflet 20., 11 pp.

SHEARER, P. S. & CULBERTSON, C. C. 1931. Outbreeding versus crossbreeding with swine. Iowa Agric. Exp. Sta. Rpt.: 24.

SHEPPERD, J. H., CHRISTENSEN, F. W., THOMPSON, O. A., & KUENNING, A. C. 1924. Hairlessness in pigs (at the Williston substation). N. Dakota Agric. Exp. Sta., Bull. 174: 96.

SILVERHJELM, W. 1933. Wyniki pracy hodowlanej w Szwecji nad polepszeniem trzody bekonowej. Przegląd hodowl., **7**: 299—300.

SIMPSON, Q. I. & J. P. 1907. Reversion induced by cross-breeding. Science, **25**: 426—428.

SIMPSON, Q. I. 1907. Rejuvenation by hybridisation. Proc. Amer. Breed. Ass.

SIMPSON, Q. I. & J. P. 1908. Genetics in swine hybrids. Science, **27**: 941.

SIMPSON, Q. I. & J. P. 1909. Inheritance of face shape in swine. Amer. Breed. Ass. Rpt., **5**: 250—255.

SIMPSON, Q. I. & J. P. 1911. Analytical hybridizing. Amer. Breed. Ass Rpt., **7**: 266—275.

SIMPSON, Q. I. 1912. Fecundity in swine. Amer. Breed. Ass. Rpt., **8**: 261—266.

SIMPSON, Q. I. 1914. Coat-pattern in mammals. J. Hered., **5**: 329—339.

SINCLAIR, R. D. & SACKVILLE, J. P. 1927. A study of some problems in bacon hog production. Alberta Coll. Agric., Bull. 15, 61 pp.

SINCLAIR, R. D. & SYROTUCK, M. 1928. Age as a factor in swine reproduction. Sci. Agric., **8**: 492—496.

SINCLAIR, R. D. 1932. Swine Production in Alberta. Alberta Coll. Agric., Bull. 22, 90 pp.

SMITH, A. D. Buchanan. 1926. Geneticist's point of view in mating pigs. Pig Breeders' Annual, **6**: 42—47.

SMITH, A. D. Buchanan & CALDER, A. 1928. Pig testing. The results of preliminary work on bacon type. Scot. J. Agric., **11**: 318—325.

SMITH, A. D. Buchanan & CALDER, A. 1930. Pig testing station. First report. 1928—30. University of Edinburgh.

SMITH, A. D. Buchanan. 1930—31. Litter size: is it inherited? Pig Breeders'
Annual, **10**: 46—52.

SMITH, A. D. Buchanan. 1933. Pig testing and recording. Agric. Prog., **10**:
114—119.

SMITH, A. D. Buchanan. 1934. University of Edinburgh, Institute of Animal Genetics. The Farm of Shothead. Edinburgh, Oliver & Boyd. 15 pp.

SMITH, G. E. & WELCH, H. 1917. Foetal athyrosis. A study of the iodine
requirement of the pregnant sow. J. Biol. Chem., **29**: 215—225.

SMITH, W. W. 1913. Colour inheritance in swine. Amer. Breed. Mag., **4**:
113—123.

SOMMER, K. 1931. Die englischen Schweinemasttypen und ihre Beurteilung
unter Berücksichtigung der Absatzverhältnisse. Arch. Tierernähr.
Tierz., **6**: 260—312.

SPENCER, S. 1921. Evolution of the improved pig. Pig Breeders' Annual,
1: 10—11.

SPILLMAN, W. J. 1906. Inheritance of coat-colour in swine. Science, **24**:
441—443.

SPILLMAN, W. J. 1907a. Colour inheritance in mammals. Science, **25**: 313—
314.

SPILLMAN, W. J. 1907b. Inheritance of the belt in Hampshire swine. Science,
25: 541—543.

SPILLMAN, W. J. 1908. Colour factors in mammals. Amer. Breed. Ass. Rpt.,
4: 357—359.

SPILLMAN, W. J. 1910. History and peculiarities of the mule-foot hog. Amer.
Breed. Mag., **1**: 178—182.

SPÖTTEL, W. 1924. Über Domestikationserscheinungen bei Schweinen. Z.
Schweinez., **32**: 65—68, 79—82, 112—114, 129—130.

STAFFE, A. 1914. Beiträge zu einer Monographie des Landschweines der
südlichen Ostalpen. Mitt. landw. Lehrkanz., Hchsch. Bodenkult., Wien,
3: 335—446.

STAHL. — 1930. Ruhlsdorfer Beobachtungen zur Ferkelaufzucht. Z.
Schweinez., **37**: 423—427.

STANG. — 1931. Schweine Versuchswirtschaft Ruhlsdorf. Berl. tierärztl.
Wschr. **47**: 709.

STEWART, W. A. 1931. Pig-keeping. Min. Agric. Fish., Bull. 32.

STRUTHERS, J. 1863. On the solid-hoofed pig; and on a case in which the
fore foot of the horse presented two toes. Edinburgh New Phil. J., **17**:
273—280.

SULTEMEIER, H. 1927. Ein Beitrag zur Leistungsprüfung in der Schweinezucht. Dtsch. landw. Presse, **54**: 297—298.

SURFACE, F. M. 1909. Fecundity of swine. Biometrika, **6**: 433—436.

SZYMANOWSKI, Z., STETKIEWICZ, St., & WACHLER, B. 1926. Les groupes
sérologiques dans le sang du porc et leur relation avec les groupes du
sang humain. C. R. Soc. Biol., **94**: 204—205.

TEODOREANU, N. I. 1922. Die Schweineborste als Rassenmerkmal. Med.
Vet. Dissertation, Hannover.

TEODOREANU, N. I. 1929. Vererbungsbeobachtungen bei Schweinen. Bull. Sect. Sci. Acad. Rouman., **12**: 26—37.

TEODOREANU, N. I. 1930. Histo-ethnologische Untersuchungen über die Struktur der Haut bei Mangalicza- und Lincolnshire-Schweinen. Bull. Sect. Sci. Acad. Rouman., **13**: 139—157.

TEODOREANU, N. I. 1931. Weitere Untersuchungen über die ethno-histologische Struktur der Haut bei verschiedenen Schweinerassen und deren Kreuzungsprodukten. Z. Zücht., B, **21**: 146—169.

TEODOREANU, N. I. 1932. Weitere Vererbungsbeobachtungen am Schwein. Bull. Sect. Sci. Acad. Rouman., **15**: 165—180.

TEXAS AGRICULTURAL EXPERIMENT STATION. 1932. Report.

THOMSON, G. M. 1922. The naturalisation of animals and plants in New Zealand. Cambridge Univ. Press, 607 pp.

TINLINE, M. J. 1922. Experiments with swine at the Scott experiment station. Canada Expt. Farms, Scott. Sta., Rpt. Supt.: 13—19.

TINLINE, M. J. 1923. Experiments with swine at the Scott experiment station. Canada Expt Farms, Scott Sta., Rpt. Supt.: 5—8.

TOOLE, W. Investigations with swine. Ontario Dept. Agric., Ann. Rpt., **47**: 33—36.

TOOLE, W. & KNOX, R. G. 1926. The bacon hog: breeding, growing and finishing. Ontario Dept. Agric., Ontario Agric. College., Bull. 320, 15 pp.

TUFF, Per. 1932. Foredlingsarbeidet i svineavlen. Norsk Landmandsblad, Nos. 5 & 6.

UHLENHUTH, P., MIESSNER, H. & GEIGER, W. 1933. Züchtung seuchenfester Schweinerassen. Dtsch. tierärztl. Wschr., **41**: 49—51.

VITZTHUM VON ECKSTAEDT, W. 1928. Fruchtbarkeitsuntersuchungen an Schweinen unter besonderer Berücksichtigung der Inzuchtverhältnisse. Züchtungskunde, **3**: 473—495.

VOLPINI, A. 1924. Della prolificità e fecondità delle scrofe. L'aveniro zoot. umbro, **4**: 6.

WALENTOWICZ, A. 1888. O przypadku dwuplciowosci obustronnej u świni. Mem. Cracow Acad. Sci., Math-nat. Sect., **14**: 71—74.

WALTHER, J. A. 1929. Cercetari asupra porcului de Basana. Bul. direct. gen. zootech. sanit. vet., **15**, 572.

WALTER, AD. R., PRÜFER, J. and CARSTENS, P. 1932. Beitrag zur Kenntnis der Vererbungserscheinungen beim Schwein. Züchter, **4**: 178—184.

WARWICK, B. L. 1926a. Inheritance of black in swine. J. Hered., **17**: 251—255.

WARWICK, B. L. 1926b. A Study of Hernia in Swine. Wisconsin Agric. Exp. Sta., Res. Bull. 69: 1—27.

WARWICK, B. L. 1927. What factors determine the number, size and weight of newborn pigs? Wisconsin Agric. Exp. Sta., Bull. 396: 64—67.

WARWICK, B. L. 1928a. Inheritance of scrotal hernia in swine. Ohio Agric. Exp. Sta., Bimonth. Bull. 13: 15—18.

WARWICK, B. L. 1928b. Prenatal growth of swine. J. Morph. Physiol., **46**: 59—84.

WARWICK, B. L. & BELL, D. S. 1930. Genetic experiments with sheep and swine. Ohio Agric. Exp. Sta., Bull. 446: 153—154.

WARWICK, B. L. 1931. Breeding experiments with sheep and swine. Ohio Agric. Exp. Sta., Bull. 480. 37 pp.

WASCHINSKY, G. 1925. Über den Thymus des Schweines. Dissertation, Berlin.

WEAVER, L. A. 1929. Swine feeds and feeding. Poland China J., 15 (12).

WELCH, H. 1917. Hairlessness and goiter in newborn domestic animals. Montana Agric. Exp. Sta., Bull. 119: 81—104.

WELCH, H. 1917. The cause and prevention of hairless pigs in the United States. Montana Agric. Exp. Sta., Circul. 71: 34—47.

WELLMANN, O. 1913. Carcass tests continued on Lincoln and Mangaliţa pigs in Hungary (t. t.). Köztelek (Budapest), 23: 3272—3275.

WELLMANN, O. 1930. Fiatal, Sertesek Taplalóanyaknukseglete. Mezögazdasági Közlony, Budapest 2 sz: 82—88.

WENCK, E. 1931. Über die Beziehungen zwischen der Entwicklung der Saugferkel und ihren späteren Mastleistungen als Grundlage der Herden und Zuchtwertbeurteilung. Z. Zücht., B. 22: 1—33.

WENTWORTH, E. N. 1912. Another sex-limited character. Science, 35: 986.

WENTWORTH, E. N. 1912. Inheritance of mammae in swine. Proc. Amer. Breed. Ass., 8: 545—549.

WENTWORTH, E. N. 1913. Inheritance of mammae in the Duroc-Jersey swine. Amer. Nat., 47: 257—278.

WENTWORTH, E. N. 1914. Sex-linked factors in the inheritance of rudimentary mammae in swine. Iowa Acad. Sci., 21: 265—268.

WENTWORTH, E. N. & AUBEL, C. E. 1916. Inheritance of fertility in swine. J. Agric. Res., 5: 1145—1160.

WENTWORTH, E. N. 1916. Large-type swine and fertility. Breeders' Gaz., 69: 722—723.

WENTWORTH, E. N. 1917. The influence of the male on litter sizes. Proc. Iowa Acad. Sci., 24: 305—308.

WENTWORTH, E. N. & LUSH, J. L. 1923. Inheritance in swine. J. Agric. Res., 23: 557—582.

WENTWORTH, E. N. 1925—26. Practical considerations in cross-breeding. Proc. Amer. Soc. Anim. Prod.: 42—49. (1927).

WENTWORTH, E. N. 1927. Some factors in the national lard trade. Monthly Letter to Animal Husbandmen: Armour's Livestock Bureau, 7, No. 11.

WERNER, J. 1897. Polydactylie beim Schweine. Sitz. Ges. naturf. Freunde Berlin: 47—48.

WESZECZKY, O. 1920. Untersuchungen über die gruppenweise Hämagglutination beim Menschen. Biochem. Z., 107: 159—171.

WHITE, G. R. 1914. Animal Castration. 227 pp., Nashville.

WIELAND, W. 1927. Ein Beitrag zur Schweinezucht auf Leistung. Z. Schweinez., 34: 798—800.

WILCKENS, M. 1886. Untersuchungen über das Geschlechtsverhältnis und die Ursachen der Geschlechtsbildung bei Haustieren. Landw. Jahrb., 15: 611—654.

WILD, H. 1927. Ergebnisse von Schweineleistungsprüfungen, insonderheit Studien über die Ferkelentwicklung nach Wiederergebnissen der Versuchswirtschaft Ruhlsdorf aus den Jahren 1923—26. Dissertation, Berlin.

WILEY, J. R. 1926. Ton litters are money makers. Chester White J., July.

WILKENS, Ch. 1929. Über Zucht und Wachstum des deutschen veredelten Landschweines im Regierungsbezirk Lüneburg.

WILLIAMS, R. H., BURNS, R. H. & SMITH, C. A. 1923. Hog-feeding investigations. Arizona Agric. Exp. Sta. Rpt.: 467—470.

WINTER, S. G. 1930. Untersuchungen über die Isohämagglutination im Blute von Pferd, Rind, Schaf, Ziege und Schwein. Dissertation, Göttingen.

WODSEDALEK, J. E. 1913. Spermatogenesis of the pig with special reference to the accessory chromosomes. Biol. Bull., **25**: 8—32.

WODSEDALEK, J. E. 1913. Accessory chromosomes in the pig. Science, **38**: 30—31.

WRIGHT, S. 1918. Colour inheritance in mammals. VIII. Swine. J. Hered., **9**: 33—38.

YATES, F. 1934. A complex pig-feeding experiment. J. Agric. Sci., **24**: 511—531.

YOUATT. 1847. See Warwick, 1926b.

ZAVORAL, H. G. 1932. Swine herd improvement projects in Minnesota. Proc. Amer. Soc. Anim. Prod.: 222—225. (1933).

ZOLLIKOFER. 1926. Leistungsprüfungen in der Schweinezucht in der Provinz Hannover.

ZORN, W. 1927. Frühreife, Futterwertung, Fruchtbarkeit, Säugefähigkeit und Wüchsigkeit der Ferkel. Dtsch. landw. Tierzucht, No. 19.

ZORN, 1931. 6. Jahresbericht der preussischen Versuchs- und Forschungsanstalt für Tierzucht in Tschechnitz. Parey, Berlin.

ZORN, W., KRALLINGER, H. F., & SCHOTT, A. 1933. Untersuchungen zur züchterischen Bewertung der Fruchtbarkeit und des Vierwochengewichtes bei weissen Edelschweinen. Züchtungskunde, **8**: 433—450.

ZWAGERMAN, C. 1930. Die Schweinehaltung in Holland. Z. Schweinez., **37**: 841—845, 861—864.

AUTHOR INDEX

SUBJECT INDEX